FILMS AND VIDEOS ON PHOTOGRAPHY

Compiled by

Program for Art on Film
Nadine Covert

Vivian Wick
Elizabeth Scheines
Teresa Abatemarco Russo

Direction des Musées de France
Gérard Turpin
Myriam Toledano

Nelly Luczak

Program for Art on Film, a joint venture between The Metropolitan Museum of Art and the J. Paul Getty Trust
Bureau de l'Audiovisuel, Direction des Musées de France, Ministère de la Culture et de la Communication

Cover photo:

MAN RAY

American, 1890 – 1976
FRAMES FROM "EMAK BAKIA," 1927
Gelatin silver print
6 7/8 x 5 7/8 in. (17.4 x 15 cm.)
The Metropolitan Museum of Art, Ford Motor Company Collection
Gift of Ford Motor Company and John C. Waddell, 1987

ISBN: 0-87099-573-1

Printed in the United States

Contents

Preface & Acknowledgments	v
How to Use the Catalogue	vii
Base Audiart	viii
The Program for Art on Film	ix
Films and Videos on Photography	1
Subject Index	85
Names Index	97
Source Index	105

Preface & Acknowledgments

The following annotated list of films and videos about photography was prepared to celebrate the 150th anniversary of the invention of photography in 1989.

It includes 511 entries, bringing together records from the Critical Inventory of Films on Art, a database compiled by the Program for Art on Film, a joint venture of The Metropolitan Museum of Art and the J. Paul Getty Trust, and from the Base Audiart, a database compiled by the Audiovisual Department of the Direction des Musées de France, Ministry of Culture and Communication, Paris. The publication has been released in both English and French editions.

Films & Videos on Photography is a result of the dedicated efforts of a number of individuals who work for the two collaborating organizations. We express our deepest thanks to the following people who worked on the compilation of this publication.

For the Program for Art on Film:

Vivian WICK, who supervised the cataloguing, indexing and editing of the entries; Elizabeth SCHEINES, who supervised data entry and copy editing of the entries and designed and implemented database formats and page layouts; Teresa ABATEMARCO RUSSO, who did research and indexing and translated French records into English; David FRIEDLANDER, senior data entry/research assistant; Susan LOGAN, indexer/researcher; Wendy WALLIS and Jeffrey GRAND, data entry/research assistants; Beatrice CARSON, copy editor and proofreader; Elizabeth CRANE, administrative assistant; Martin FRITZ, researcher/translator (German).

For Base Audiart:

Myriam TOLEDANO, who supervised the cataloguing, indexing, and editing of the entries, did research, and translated English records into French; Nelly LUCZAK, data entry/researcher; Muriel BARAN and Patrice BELLEAU, typists; Valérie LAVERGNE-DUREY, researcher/data entry.

We also wish to express our appreciation to the following for their advice and assistance:

Miles BARTH, International Center of Photography; Toni PETERSEN and Cathleen WHITEHEAD, Art and Architecture Thesaurus, Getty Art History Information Program; Jan-Christopher HORAK, George Eastman House; Museum of Modern Art Film and Photography Study Centers; and the staff of the Editorial and Publishing divisions of The Metropolitan Museum of Art, especially Barbara BURN, Lisa KOCH, Valerie TROYANSKY, David KROL, Janice HERRMAN, and Linda CURLEY.

Anne COUTINOT, Intermédia; André FAGE, Musée Français de la Photographie; Alain SAYAG, Musée National d'Art Moderne, Centre Georges Pompidou; Monique DE LINARES, Chantal SOYER, Mission pour l'Audiovisuel; Pierre BONHOMME, Mission du Patrimoine Photographique; Elizabeth PUJOL, Centre National de la Photographie; Françoise WASSERMAN, Ecomusée de Fresnes; Christine LOISY, Diffusion des Productions Audiovisuelles du Musée d'Orsay.

Nadine Covert
Critical Inventory
Program for Art on Film

Gérard Turpin
Audiovisuel
Direction des Musées de France

How to Use the Catalogue

The documentary and educational films included in this catalogue are listed alphabetically by title. The appendices include a Subject Index, which will enable users to identify films on specific topics. Titles may be listed under more than one subject heading. Works on particular photographers can be found by consulting the Names Index. The Source Index lists addresses of distributors or producers cited in each entry.

Each entry includes filmographic data, credits, source, and a brief annotation. The following example explains the various elements:

Title	**REPORTERS**
Running Time; Color/Black & White; Format; Date of Release	97 min. col. 16mm 1981
Country of production; Language	France French with English subtitles
Director	**Dir.:** Raymond Depardon
Producing Agency	**Prod. Agency:** Centre Georges Pompidou; Pari Films
U.S. Distribution	**Distributor (US):** Cinema Guild **Distributor (US):** Interama
Foreign Distribution	**Foreign Distr:** Pari Films (FR) **Foreign Distr:** Cinéculture (FR)
Synopsis	French photographer and filmmaker Raymond Depardon (b.1942) looks at the lives of Parisian photojournalists and presents a sympathetic view of the hard-working, unyielding photographers. Depardon followed reporters from the Gamma Agency for the month of October 1980 to record the daily events of the Parisian news scene. Stars some of the leading players in French political and artistic life: François Mitterrand, Jean-Luc Godard, Valéry Giscard d'Estaing, Catherine Deneuve, Jacques Chirac, Alain Delon, Georges Marchais, and Mireille Mathieu. There are guest appearances by visitors to Paris, such as President Bongo of Gabon and American actors Gene Kelly and Richard Gere. (*Cinema Guild*)
Reviews	**Reviews:** *Variety*; *New York Post*; *New York Times*; *Arts Weekly*; *Village Voice*, 21 Oct 86
Awards	**Awards:** César Best Documentary, 1981

Base Audiart

Database of Films on Art

The audiovisual office of the Direction des Musées de France has created a world inventory of film and video programs on art. A software program was written in collaboration with the Ministry of Culture to set up a computer databank, Base Audiart, which now includes more than 10,300 entries.

The films and videotapes inventoried in this databank were grouped into seventeen different categories:

Archaeology, Architecture, Art History, Civilization, Crafts, Ethnography, Decorative Arts, Fine Arts, Graphic Arts, Painting, Photography, Prehistory, Public Institutions, Science in the Service of Art, Sculpture, Techniques, Video art.

The Base Audiart responds mainly to the following questions: Which films deal with what subjects? Where can one find them?

Information in the various fields may be accessed in a number of different ways: by title of film, dates of production, names of producers, distributors, directors, artists, geographical areas, and so on. This categorization assures optimal service for cultural institutions (museums, galleries, educational establishments, and libraries) and for those in the audiovisual professions.

To request information, contact the audiovisual office at the address below, attention Myriam Toledano or Nelly Luczak.

Catalogues, brochures and information about films and videotapes can be sent to:

DIRECTIONS DES MUSEES DE FRANCE
Bureau des Affaires Scientifiques
Informatiques et Audiovisuelles
34 Quai du Louvre - 75041 Paris CEDEX 01
Tel. 42.60.39.26 - ext. 3659 et 3707

The Program For Art on Film

The Program for Art on Film is a joint venture of The Metropolitan Museum of Art, New York, and the J. Paul Getty Trust, Los Angeles, affiliated with the Metropolitan Museum's Office of Film and Television. Established in 1984, the Program has as its goals the improvement of the quality of film and video productions about art and the enhancement of public awareness of the visual arts through the media of film, video and television.

The first project undertaken by the joint venture was the Critical Inventory of Films on Art, an international compilation of information about films and videotapes on the visual arts (fine arts, architecture, archaeology, photography, decorative arts, and crafts). The Inventory is maintained on computer and provides a unique central repository of information that can be accessed in a variety of ways: by title, date, credits, country of production, and so on, as well as by subject matter. The Inventory includes data on some 15,000 films and videos. Current productions from the United States and overseas are evaluated by experts in art history, art education, film and television. These critical evaluations are incorporated into the database to assist film users in selecting the most appropriate films for their programs. Selected portions of the database will be published in a directory of films on art as well as in filmographies on special topics.

In late 1986 the Program for Art on Film initiated a Production Laboratory. Fourteen short film or video productions will have been completed by 1990, all of them involving a collaboration between an art historian or artist and a film director, and all of them focusing on an aspect of art before Impressionism. Each of the productions had a "brief" — an experiment they were attempting, a problem they sought to resolve, a filmic challenge they undertook in order to explore innovative or exemplary approaches to presenting art on film.

The Executive Director of the Program is Karl Katz, Consultant for Film and Television at the Metropolitan Museum. Dr. Wendy A. Stein, Program Manager, is in charge of administration. Nadine Covert is the Special Consultant in charge of the Critical Inventory. Joan Shigekawa is Coordinating Producer for the Production Laboratory.

Producers and distributors are encouraged to send information about their productions to the attention of Nadine Covert for inclusion in the Critical Inventory.

For more information, contact: Program for Art on Film
980 Madison Ave.
New York, NY 10021
Tel. 212-988-4876
FAX 212-628-8963

FILMS AND VIDEOS ON PHOTOGRAPHY

A. DUMAS OU L'HYPNOSE CINEMA

13 min. b&w 16mm 1981
France French
Foreign Distr: Intermédia (FR)
Foreign Distr: Association pour la Diffusion de la Pensée Française (FR)

French author Alexandre Dumas (1824-1895) and other writers were interested in hypnosis and in chronophotography, a method of time/motion photography invented by French physiologist, inventor, and photographer Jules Etienne Marey (1830-1904). The writers used these two techniques to produce works of a disturbing strangeness.

A.D. COLEMAN

Series Title: On Art and Artists
28 min. b&w video 1977
United States English
Prod. Agency: Video Data Bank
Distributor (US): Video Data Bank

Alex Sweetman interviews American photography critic A.D. Coleman (b.1943), who discusses his role and stance as a critic and the issues crucial to the development of photography criticism. On Art and Artists series.

AARON SISKIND

17 min. col. 16mm; video 1981
United States English
Dir.: Theodore R. Haimes **Prod.:** Edgar B. Howard
Prod. Agency: Checkerboard Foundation, Inc.
Distributor (US): artsAmerica
Distributor (US): American Federation of Arts
Distributor (US): International Center of Photography (Archival only)
Distributor (US): Rental-see *Educational Film & Video Locator*

Profile of American photographer Aaron Siskind (b.1903), who discusses his aesthetic philosophy and describes his transition in 1943-1944 from a documentary style to a more personal, abstract one. Filmed on location in Lima, Peru, where Siskind is shown photographing his favorite subject — walls with abstract remnants of posters and graffiti — and on Martha's Vineyard, Massachusetts, where specific sites inspired his abstract style.
Reviews: *EFLA Evaluations*, Spring 84; *New York Times*, 4 Sept 85

ACTION PHOTOGRAPHY

Series Title: Exploring Photography
30 min. col. video 1986
United States English
Prod. Agency: Eastman Kodak Co.
Distributor (US): Baker & Taylor Video (Video sales only)
Distributor (US): Aylmer Press (Video sales only)
Distributor (US): Kodak Video Programs

Photographer John Vaeth demonstrates how to make action pictures of sports events using a 35mm camera with interchangeable lenses and high-speed film. Discusses basic equipment and the effects of various lenses (wide angle, 50mm, telephoto) and filters.Shows how to prefocus, follow and stop action, pan the camera, get good indoor shots with high-speed film, and capture the "peak moment" of action. Exploring Photography series.
Reviews: *Landers Film Reviews*, Vol 32, No 1, Fall 87

ALBEDO

54 min. col. 16mm 1982
Canada French with English subtitles
Dir.: Jacques Leduc; René Roy **Prod.:** Jacques Bobet
Prod. Agency: National Film Board of Canada
Foreign Distr: National Film Board of Canada (CAN)

Based on the journals and photographs of David Marvin of Montreal, chronicles the life of the photographer who, tormented by deafness and the hardships of living, killed himself in 1975. Surrealistic in style, the film parallels Marvin's life with that of a dying Montreal neighborhood, Griffintown, which he faithfully documented.

ALBERT TESSIER A FORCE D'IMAGES

58 min. col. 16mm 1976
Canada French
Dir.: André Ricard **Prod.:** Pauline Geoffrion
Prod. Agency: Film Cenatos; National Film Board of Canada
Foreign Distr: CEGEP de Trois-Rivières (CAN)
Foreign Distr: National Film Board of Canada (CAN)

Evokes the unique memories of Canadian photographer Albert Tessier (1895-1976).

ALFRED EISENSTAEDT

Series Title: Master Photographers; No. 1
35 min. col. 16mm 1982
Great Britain English
Prod.: Peter Adam
Prod. Agency: BBC-TV; MM Chanowski Productions bv
Foreign Distr: Arts Council of Great Britain Film Library (GB)

Profiles German-born American photographer Alfred Eisenstaedt (b.1898), who has been called the father of photojournalism. His work is a catalogue of people and events that shaped the twentieth century. Shows Eisenstaedt in his New York office, where he discusses some of his portraits, including those of Richard Strauss, Arturo Toscanini, Marilyn Monroe, Winston Churchill, and Adolf Hitler. Master Photographers series, No. 1 of 6.

ALFRED STIEGLITZ: PHOTOGRAPHER (1864-1946)

26 min. col.& b&w 16mm 1982
United States English
Dir.: Hans Namuth; Paul Falkenberg **Prod.:** Paul Falkenberg; Hans Namuth
Prod. Agency: Museum at Large, Ltd.
Distributor (US): Museum at Large, Ltd.
Distributor (US): American Federation of Arts
Distributor (US): International Center of Photography (Archival only)

Profile of pioneering American photographer Alfred Stieglitz (1864-1946), editor of *Camera Work* and patron of artists such as John Marin, Arthur Dove, and Georgia O'Keeffe, who was also Stieglitz's wife. He was one of the foremost figures in twentieth-century American art. Shows Stieglitz himself, his photographs, and the works of artists he encouraged. Narrated by art critic John Russell with reminiscences by Ansel Adams, Mary Steichen, Isamu Noguchi, Dorothy Norman, Mary Rapp, Aaron Copland, and Arnold Newman.
Reviews: *EFLA Evaluations*, Fall-Winter 83/84
Awards: American Film Festival Red Ribbon, 1983

AMERICA AND LEWIS HINE

58 min. col.& b&w 16mm; video 1984
United States English
Dir.: Nina Rosenblum **Prod.:** Nina Rosenblum; Daniel V. Allentuck
Exec. Prod.: David Loxton
Prod. Agency: Daedalus Productions; WNET/Channel 13
Distributor (US): Cinema Guild
Distributor (US): Daedalus Productions

Biographical documentary of the life, work, and times of America's pioneer industrial photographer and social activist Lewis Hine (1874-1940), who recorded the development and adverse effects of the Industrial Revolution in the United States from 1900 to 1940. Mixes Hine's photographs, historical footage, handbills, book-cover reproductions, interviews, reminiscences, and excerpts from Hine's letters, field reports, and articles to convey his passionate dedication to documenting and changing his times. Juxtaposes vintage footage of New York cityscapes, West Virginia coal mines, and southern textile factories with images of newly arrived immigrants, sweatshop employees, homebound pieceworkers, bedraggled child laborers, and others toiling in undesirable conditions. Hine died virtually unknown and penniless in 1940, but his works are now recognized as treasures of America's national heritage. Musical score features folk songs and period music. Includes interviews with Anthony "Shorty" Slick, an eighty-year-old coal miner who worked as a breaker boy; Harold McClain, the last surviving ironworker who helped build the Empire State building in New York City, a project for which Hine was official photographer; and photographer Walter Rosenblum, who befriended Hine at the New York Film and Photo League in 1938 and curated the

posthumous exhibition and publication of his photographs. Narrated by John Crowley, with Jason Robards as the voice of Lewis Hine and Maureen Stapleton as the voice of social reformer Margaret Byington.
Reviews: *Booklist*, Vol 81, No 22; *Booklist*, Vol 82, No 10; *Library Journal*, Vol 110, No 17; *Landers Film Reviews*, Vol 30, No 1; *Variety*, 29 Aug 84; *New York Times*; *Ms. Magazine*
Awards: American Film Festival Red Ribbon, 1985; Montreal Festival Intl. du Film sur l'Art Best Biography of an Artist, 1985; CINE Golden Eagle, 1985; Baltimore Intl. Film Festival First Prize; U.S. Film Festival Special Jury Prize; Sinking Creek Film Festival Cash Prize

AMERICAN IMAGE: 150 YEARS OF PHOTOGRAPHY

60 min. col.& b&w video 1988
United States English
Dir.: Bob Comiskey **Prod.:** Debra Shapiro
Exec. Prod.: Martin Sandler; Liz Chang; Dick Amaral
Prod. Agency: Martin Sandler Productions; Eastman Kodak Co.; WCBS-TV
Distributor (US): Kodak Video Programs
Distributor (US): Aylmer Press (Video sales only)
Distributor (US): Facets Multimedia (Video sales only)
Distributor (US): Baker & Taylor Video (Video sales only)

Surveys the development of photography and its impact on daily life from the image-etching process discovered by French painter and photographer Louis Jacques Mandé Daguerre (1787-1851) to the present. Traces the history of picture taking in the United States from early tintype portraits to satellite photography. Discusses photography as an art form and examines the work of photographers Gertrude Käsebier, Alfred Stieglitz, Clarence H. White, and Ansel Adams.
Awards: CINE Golden Eagle, 1989

AMERICAN IMAGES: PHOTOGRAPHS AND PHOTOGRAPHERS OF THE FARM SECURITY ADMINISTRATION, 1932-1942

60 min. col.& b&w video 1980
United States English
Prod. Agency: Amarillo Art Center
Distributor (US): Amarillo Art Center

Documents the 1979 reunion held in Amarillo, Texas, of photographers who worked for the Farm Security Administration recording American farm life during the years 1932 to 1942. Shows many photographs, including some by Dorothea Lange (1895-1965) and Walker Evans (1903-1975). Accompanied by music of the period by Woody Guthrie and Fletcher Henderson.

ANDERSEN CHEZ LE PHOTOGRAPHE

13 min. col. 16mm 1975
Denmark Danish?
Dir.: Jorgens Roos; Henrik Poulsen
Prod. Agency: Statens Filmcentral
Foreign Distr: Ambassade du Danemark (FR)

The many photographs of Danish writer Hans Christian Andersen (1805-1875) give a visual impression of the great storyteller and the art of photography at the end of the nineteenth century.

ANDRE KERTESZ

Series Title: Master Photographers; No. 5
35 min. col. 16mm 1983
Great Britain English
Prod.: Peter Adam
Prod. Agency: BBC-TV; MM Chanowski Productions bv; Andreas Landshoff Productions
Foreign Distr: Arts Council of Great Britain Film Library (GB)

Televised interview with Hungarian-born American photographer André Kertész (1894-1985), who talks about his early life in Hungary, Paris in the 1920s, and his long, bitter battle for acceptance in the United States. Master Photographers series, No. 5 of 6.

ANDRE KERTESZ OF THE CITIES (ANDRE DANS LES VILLES: BUDAPEST, PARIS, NEW YORK, 1894-1985: PORTRAIT D'ANDRE KERTESZ)

55 min. col.& b&w 16mm; video 1985
France French
Dir.: Teri Wehn Damisch **Prod.:** Teri Wehn Damisch
Exec. Prod.: Gérard Barralis; Jean-Emile Jeannesson
Prod. Agency: TF 1 with the participation of the Ministère de la Culture
Foreign Distr: Centre National des Arts Plastiques (FR)
Foreign Distr: TF 1 (FR)

A portrait of Hungarian-born American photographer André Kertész (1894-1985), filmed shortly before his death. Follows Kertész as he revisits the places of his youth and early career in Budapest, Paris, and finally New York, where he arrived in 1936. Kertész comments on his personal history and his work as he reshoots famous photographs he made decades before. He is seen at work in his studio, experimenting with new images and preparing for his next exhibition. For the film, he repeats his famous experiments shooting nudes in a distorting mirror. Made in conjunction with the exhibition "André Kertész: Of

Paris and New York," organized by the Art Institute of Chicago in collaboration with The Metropolitan Museum of Art, New York City.

ANDRE KERTESZ, PHOTOGRAPHE AMERICAIN A PARIS: 1981

Series Title: Aujourd'hui en France; No. 45
5 min. col. 16mm; video 1981
France French (English, Spanish, and Arabic language versions also avail.)
Prod. Agency: French Ministry of Foreign Affairs
Distributor (US): Embassy of France
Foreign Distr: French Ministry of Foreign Affairs (FR)
Foreign Distr: ICS (FR)

Paris proclaimed November 1980 "The Month of the Photo." It was an opportunity to celebrate the work of André Kertész (1894-1985), the most French of American photographers. Aujourd'hui en France series, No. 45.

ANDRE KERTESZ: A POET WITH A CAMERA

29 min. col. video 1985
United States English
Dir.: Bela Urgin **Prod.:** Bela Urgin
Distributor (US): International Center of Photography (Archival only)

Presents the life and work of Hungarian-born American photographer André Kertész (1894-1985). Cornell Capa, founding director of the International Center of Photography in New York City, comments on Kertész's work. Includes footage of Kertész at home and at work in his darkroom.

ANDREAS FEININGER

Series Title: In and Out of Focus
30 min. col. video 1978
United States English
Prod.: Casey Allen
Prod. Agency: WNYC
Distributor (US): WNYC
Distributor (US): Center for Creative Photography
Distributor (US): International Center of Photography (Archival only)

American photographer Andreas Feininger (b.1906) discusses his book *The Mountains of the Mind*. In and Out of Focus series.

ANDREAS FEININGER

Series Title: Master Photographers; No. 3
35 min. col. 16mm 1982
Great Britain English
Prod.: Peter Adam
Prod. Agency: BBC-TV; MM Chanowski Productions bv
Distributor (US): International Center of Photography (Archival only)
Foreign Distr: Arts Council of Great Britain Film Library (GB)

Profiles American photographer Andreas Feininger (b.1906), who was an architect before he took up photography. The urban landscape remains one of his major themes. Focuses on the technical possibilities of creative photography and Feininger's belief that the camera can enrich a person's way of seeing. Master Photographers series, No. 3 of 6.

ANDREW J. RUSSELL, A VISUAL HISTORIAN

28 min. col. 16mm; video 1983
United States English
Dir.: James W. Dearden
Prod. Agency: Brigham Young University, Media Productions
Distributor (US): Britannica Films
Distributor (US): Brigham Young University, Audio Visual Services

A study of the photographic documentation of American history, people, geography, and events made by Andrew J. Russell (1830-1902). Russell's biographer, historians, a photographic archivist, and a locomotive engineer comment on his work. Also includes a demonstration of the laborious process of taking photographs in Russell's day.

ANIMALS IN MOTION

7 min. col. 16mm 1968
Canada Nonverbal
Dir.: John Straiton **Prod.:** John Straiton
Distributor (US): Biograph Entertainment, Ltd.
Foreign Distr: Canadian Filmmakers Distribution Centre (CAN)

Animated homage to the British photographer Eadweard Muybridge (1830-1904), a motion picture pioneer who conducted exhaustive experiments in sequential photography, documenting the components of movement.

Samples of his sepia stills are compiled into a compendium of unattired Victorian men, women, and children dancing, boxing, exercising, or just cavorting. Nude humans are compared to nude animals in many variations of work and play.
Awards: Canadian Film Awards Etrog, 1975; Cannes Amateur Film Festival; Australian Amateur Cine Society

ANNEES KAGAN, LES
29 min. col. video 1989
France French
Dir.: Jean Pierre Krief
Prod. Agency: La Sept; KS Visions; SGGC; PHAZ
Foreign Distr: KS Visions (FR)
Foreign Distr: La Sept (FR)
Profiles French photojournalist Elie Kagan, who works independently with no affiliation to any press agency. Artisan, rebel, and archivist, he photographs social and political events, such as the Poujadist demonstrations of 1956 and the manifestations of 1968.

ANNEES-DECLIC, LES
70 min. b&w 35mm; 16mm 1983
France French
Dir.: Raymond Depardon; Roger Ikhlef
Prod. Agency: I.N.A.; Centre National de la Photographie
Distributor (US): Interama
Foreign Distr: Intermédia (FR)
Foreign Distr: Pari Films (FR)
French photographer and filmmaker Raymond Depardon (b.1942) contemplates his life and work over a twenty-year period. Presents a retrospective sampling of his photographs from 1957 to 1977 and excerpts from his documentary films to retrace his career as a photojournalist. He talks about his coverage of daily events of Parisian life and the dangerous times he spent in the African desert photographing skirmishes in Chad and Algeria.
Reviews: *Boston Globe*, 14 Oct 87

ANSEL ADAMS
Series Title: Master Photographers; No. 6
35 min. col. 16mm 1982
Great Britain English
Prod.: Peter Adam
Prod. Agency: BBC-TV; MM Chanowski Productions bv
Foreign Distr: Arts Council of Great Britain Film Library (GB)
Profiles American landscape photographer Ansel Adams (1902-1984). Shows Adams capturing the rugged grandeur of California, which inspired so much of his work. Presents a selection of his photographs, many of which now hang in museums. Master Photographers series, No. 6 of 6.

ANSEL ADAMS: PHOTOGRAPHER
60 min. col. 16mm; video 1981
United States English
Dir.: John Huszar **Prod.:** John Huszar; Andrea Gray
Prod. Agency: Film America, Inc.
Distributor (US): Arthur Cantor, Inc.
Distributor (US): Film America, Inc.
Distributor (US): artsAmerica
Distributor (US): Facets Multimedia (Video only)
Distributor (US): Pacific Arts Video (Video sales only)
Distributor (US): University of Illinois Film Center
American photographer Ansel Adams (1902-1984) speaks about his long and prolific career as an artist, conservationist, and teacher. Follows him to the locations of his most famous photographs. Includes on-camera conversations with artist Georgia O'Keeffe and photographic historian Beaumont Newhall. Adams discusses theoretical and emotional aspects of photography and his own photographs in particular, using many of his best-known works as examples. He also outlines the development of photography as an art form and his part in the creation of the Department of Photography at the Museum of Modern Art, New York City.
Reviews: *EFLA Evaluations*, May-June 83; *Artnews*, Vol 81, No 6, Summer 88; *Library Journal*, Aug 87
Awards: CINE Golden Eagle, 1981

ANSEL ADAMS: PHOTOGRAPHER
20 min. b&w 16mm; video 1957
United States English
Dir.: David Myers **Prod.:** Larry Dawson
Prod. Agency: Larry Dawson Productions
Distributor (US): International Film Bureau
Distributor (US): International Center of Photography (Archival only)
Distributor (US): Rental-see *Educational Film & Video Locator*
Foreign Distr: Educational Film Distributors (CAN)
Profile of Ansel Adams (1902-1984), the American photographer who was also a musician, mountaineer, writer, and philosopher. Describes his equipment and shows how

he used his zone system of control over exposure and development. Explores his photography of America's parks and monuments and touches briefly on his work as a commercial photographer, including his early work for Polaroid. He talks about the development of photography as an art form and discusses his attitude toward art in general and music in particular. He is shown teaching his annual photography workshop. Shows many photographs, including his portrait studies of Edward Weston and Alfred Stieglitz, and the famous *Moonrise, Hernandez, New Mexico, 1941*. Narrated by photographic historian Beaumont Newhall.

ANSEL ADAMS: PHOTOGRAPHER OF THE AMERICAN WEST

9 min. col. 16mm 1981
United States English
Dir.: Barbara Anne Richards
Prod. Agency: International Communication Agency
Distributor (US): United States Information Agency (Overseas use only)

Interviews American photographer Ansel Adams (1902-1984) at his home. Adams discusses his beginnings in photography, the changes in his style, and his role as a teacher.

APPAREIL PHOTOGRAPHIQUE, L'

26 min. col. 16mm 1984
France French
Dir.: P. Gauthier; D. Peray
Foreign Distr: Cinémathèque de l'Enseignement (FR)

Students use documents and examples to research photographic equipment. Examines the construction of a darkroom and techniques for its use with or without lenses.

ARNULF RAINER

Series Title: Portraits d'Artistes
43 min. col. 16mm 1983
Luxembourg German
Dir.: Liliane Thorn Petit **Prod.:** Liliane Thorn Petit
Prod. Agency: Radio-Télé-Luxembourg
Foreign Distr: Westdeutscher Rundfunk (BRD)

Profiles Austrian painter, sculptor, and photographer Arnulf Rainer (b.1929). Portraits d'Artistes series.

ARNULF RAINER

44 min. col. 16mm 1976
Federal Republic of Germany German
Dir.: Wibke von Bonin
Prod. Agency: Westdeutscher Rundfunk
Foreign Distr: Westdeutscher Rundfunk (BRD)

Profiles Austrian painter, sculptor, and photographer Arnulf Rainer (b.1929).

ARNULF RAINER

40 min. b&w 16mm 1970
Austria German
Dir.: Herbert Brodle
Prod. Agency: ORF
Foreign Distr: ORF (AUST)

Profiles Austrian painter, sculptor, and photographer Arnulf Rainer (b.1929).

ART ET LA DECADENCE, L'

Series Title: Haute Curiosité
44 min. col. 16mm 1987
France French
Dir.: Jean Aurel **Prod.:** Maurice Rheims
Prod. Agency: FR3
Foreign Distr: FR3 (FR)

Ponders the meaning of decadence and whether the end of the nineteenth century was decadent or triumphant. Discusses the myth of the decadence of the Roman Empire, the fall of the Austrian Empire, and decadent photography. Haute Curiosité series.

ART ET L'ILLUSION, L'

Series Title: Haute Curiosité
50 min. col. 16mm 1987
France French
Dir.: Jean Aurel **Prod.:** Maurice Rheims
Prod. Agency: FR3
Foreign Distr: FR3 (FR)

Discusses the idea that the imitation of nature or rendering the illusion of reality has been and often still is the avowed goal of artists. Includes the participation of Australian photographer Helmut Newton (b.1920), anthropologist Claude Levi-Strauss, Serge Bramley, Bettina Rheims, and Gerard Majax. Haute Curiosité series.

ART IN AMERICA, PART 10: PHOTOGRAPHY, MASTERS OF THE TWENTIETH CENTURY

Series Title: Americana Series, The
25 min. col. 16mm; video 1981

United States English
Dir.: Irene Zmurkevych **Prod.:** Irene Zmurkevych
Exec. Prod.: Leo A. Handel
Prod. Agency: Handel Film Corp.
Distributor (US): Handel Film Corp.
Distributor (US): University of Illinois Film Center (Rental only)
Distributor (US): Syracuse University Film Rental Center (Rental only)

Deals with the efforts of inventors and artists to perfect the mechanical aspects of photography that affect the potential for aesthetic expression. Shows images which capture moments of historical, social, and environmental significance and which reflect the artistic vision of the photographers. Includes interviews with several pioneers of twentieth-century photography. Art in America series, Part 10 of 10; from The Americana Series.
Reviews: *Landers Film Reviews*, Mar-Apr 82; *Arts & Activities*, Dec 82
Awards: CINE Golden Eagle, 1981

ART IN AMERICA, PART 9: PHOTOGRAPHY, THE BEGINNING OF A NEW ART

Series Title: Americana Series, The; No. 31
25 min. col. 16mm; video 1981
United States English
Dir.: Irene Zmurkevych **Prod.:** Irene Zmurkevych
Exec. Prod.: Leo A. Handel
Prod. Agency: Handel Film Corp.
Distributor (US): Handel Film Corp.
Distributor (US): University of Illinois Film Center (Rental only)
Distributor (US): Syracuse University Film Rental Center (Rental only)

Traces the history of artistic photography from the time of the invention of the daguerreotype in 1839. Art in America series, Part 9 of 10; from The Americana Series, No. 31.
Reviews: *Landers Film Reviews*, Sept-Oct 81; *Arts & Activities*, Sept 81

ART OF MARIE COSINDAS, THE

29 min. col. 16mm 1967
United States English
Prod. Agency: NET
Distributor (US): Indiana University Audio Visual Center
Distributor (US): Pennsylvania State University Audio Visual Services
Distributor (US): Rental-see *Educational Film & Video Locator*

Presents the work of American photographer Marie Cosindas (b.1925) along with comments by museum visitors, art critics, and former subjects. Shows her discussing her work with others, creating a still life, and working on two portraits. Includes a clip of one of her experimental films. (*Educational Film &Video Locator*)

ATELIER DE CHRISTIAN BOLTANSKI, L': 1983

Series Title: Aujourd'hui en France; No. 113
3 min. col. 16mm; video 1983
France French (English, Spanish, and Arabic language versions also avail.)
Prod. Agency: French Ministry of Foreign Affairs
Distributor (US): Embassy of France
Foreign Distr: French Ministry of Foreign Affairs (FR)
Foreign Distr: ICS (FR)

Features French conceptual artist and photographer Christian Boltanski (b.1944), whose works use toys and photography. Aujourd'hui en France series, No. 113.

ATGET

30 min. b&w 16mm 1967
United States English
Prod. Agency: NET
Distributor (US): Indiana University Audio Visual Center
Distributor (US): Rental-see *Educational Film & Video Locator*

Presents works by French photographer Eugène Atget (1857-1927), particularly celebrated for his images of Paris. Narration by American photographer Berenice Abbott (b.1898), who acquired and preserved his prints and negatives after his death.
Reviews: *Sightlines*, Vol 16, No 1 & 2, Fall-Winter 82/83

ATHLETE, THE: ROBERT RIGER'S DRAWINGS, PHOTOGRAPHS AND TELEVISION SPORTS

28 min. col. video 1982
United States English
Dir.: Andrew Silver; Yong Hee Silver
Prod.: Andrew Silver; Yong Hee Silver
Prod. Agency: Silver Productions
Distributor (US): Silver Productions
Distributor (US): Video Gallery
Distributor (US): Guidance Associates

Explores the works of American sports photographer, filmmaker, and graphic artist Robert Riger. Includes interviews with the artist and footage from his films. Accompanied by Riger's commentary, exhibits thirty years of the artist's work, culled from his sixteen books and his personal collections. Riger discusses his years as a producer for ABC Sports.

AU PAYS DES VISAGES

45 min. col. 16mm 1972
France French
Dir.: Frédéric Rossif
Prod. Agency: Télé Hachette; Bayerischer Rundfunk
Foreign Distr: UGC Droits Audiovisuels (FR)

Features portraits of famous artists and writers by German-born French photographer Gisèle Freund (b.1912), accompanied by excerpts from their writings about their life and art.

AUSCHWITZ L'ALBUM LA MEMOIRE

42 min. col.& b&w video 1983
France French
Dir.: Alain Jaubert
Prod. Agency: I.N.A.; C.N.R.S.; Hexagramm
Foreign Distr: Delta Image (FR)

Shows an album of 189 photographs taken by an SS photographer of the arrival of Hungarian Jews at Auschwitz, Poland, in 1944. Four French women, who were deported to Auschwitz, comment on the images in light of their own memories.

AUTOCHROMES, LES

Series Title: Zig Zag
26 min. col. 16mm 1981
France French
Dir.: Pierre Desfons **Prod.:** Teri Wehn Damisch
Prod. Agency: Antenne 2
Foreign Distr: Antenne 2 (FR)

Presents early color photographs. Zig Zag series.

AUTOUR DE LA PHOTOGRAPHIE

15 min. col. 16mm 1978
France French
Dir.: Patrice Gauthier
Foreign Distr: Cinémathèque de l'Enseignement (FR)

Describes the problems of light and the idea and quality of the image. Introduces students to photographic equipment.

BANDE A MAN RAY, LA

Series Title: Heures Chaudes de Montparnasse, Les; No. 6
50 min. b&w 16mm 1964
France French
Dir.: Jean Marie Drot **Prod.:** Jean Marie Drot
Prod. Agency: FR3; INA; La Sept
Foreign Distr: INA (FR)

Presents photographs of the surrealists by American painter, photographer, and filmmaker Man Ray (1890-1976). Les Heures Chaudes de Montparnasse series, No. 6 of 7.

BANLIEUES ENCEINTES

26 min. col. video 1986
France French
Dir.: François Gaulier
Foreign Distr: Heure Exquise Distribution (FR)

Presents several photographic experiments.

BARBARA MORGAN

Series Title: Photoprofiles
30 min. col. video 1983
United States English
Dir.: John Morgan; Rebecca L. Morgan; Paul R. Schranz; Margaretta K. Mitchell **Prod.:** Thomas R. Schiff
Prod. Agency: Images Productions
Distributor (US): Images Productions

Profiles American photographer Barbara Morgan (b.1900), who discusses her photographs of dancer and choreographer Martha Graham and her early work as a painter. In addition, she recalls her collaboration with her husband, Willard D. Morgan, especially their trips to the American Southwest, where they photographed and drew its surreal landscape. Photoprofiles series.

BARBARA MORGAN: EVERYTHING IS DANCING

18 min. col. video 1983
United States English
Dir.: Theodore R. Haimes **Prod.:** Edgar B. Howard
Prod. Agency: Checkerboard Foundation, Inc.
Distributor (US): American Federation of Arts
Distributor (US): International Center of Photography (Archival only)

American photographer Barbara Morgan (b.1900) explains why dance and movement became important subjects for her work. Discusses her photographs of dancer and choreographer Martha Graham and her dance company taken in the 1930s.

BEHIND THE BIG PICTURE

15 min. col. video 1986
United States English
Dir.: Eric Shambroom **Prod.:** Eric Shambroom
Distributor (US): DeCordova Museum and Sculpture Park

Produced in conjunction with "The Big Picture," a traveling exhibition consisting of the Polaroid Collection of original 20x24 color prints, which was shown at the DeCordova Museum and Sculpture Park, Lincoln, Massachusetts. Features Barbara Hitchcock, curator of the Polaroid Collection, and American photographers William Wegman, Chris Enos, and Olivia Parker. John Reuter of Polaroid explains the virtues of the film and camera.

BENNO FRIEDMAN

28 min. col. video 1977
United States English
Prod. Agency: Video Data Bank
Distributor (US): Video Data Bank

Alex Sweetman interviews American photographer Benno Friedman (b.1945), who discusses his work and his use of bleaching, staining, hand coloring, and drawing to manipulate the photographic image. The artist talks about the expediency and mobility of photography compared with sculpture and painting, media which he has abandoned.

BERNARD FAUCON

Series Title: Inaperçu, L'
20 min. col. 16mm 1986
France French
Dir.: Jean Claude Larrieu **Prod.:** Thierry Garrel
Prod. Agency: INA
Foreign Distr: Antenne 2 (FR)

Portrait of French conceptual artist and fashion photographer Bernard Faucon (b.1950). Discusses his models, mannequins, and use of special effects. L'Inaperçu series.

BERNARD FAUCON, PHOTOGRAPHE

Series Title: Autoportrait
13 min. col. 16mm 1988
France French
Dir.: Jean Real
Prod. Agency: Cinq Continents
Foreign Distr: Cinq Continents (FR)

A collaboration between French conceptual artist and photographer Bernard Faucon (b.1950) and filmmaker Jean Real in which the artist conceives the sequences, provides visual or audio elements, and holds the film camera. Without analyzing Faucon's photographs, presents the universe of imagery contained in his work. Autoportrait series.

BERNARD FAUCON, PHOTOGRAPHE

Series Title: Aujourd'hui en France; No. 151
4 min. col. 16mm; video 1985
France French (English, Spanish, and Arabic language versions also avail.)
Prod. Agency: French Ministry of Foreign Affairs
Distributor (US): Embassy of France
Foreign Distr: French Ministry of Foreign Affairs (FR)
Foreign Distr: ICS (FR)

French conceptual artist and photographer Bernard Faucon (b.1950) fashions his photos like paintings. He stages his landscapes and models and photographs only in color. Aujourd'hui en France series, No. 151.

BERT HARDY

30 min. col. video 1985
Great Britain English
Dir.: Noel Chanan; Suzanne Davies **Prod.:** Suzanne Davies
Prod. Agency: Joseph Plateau Productions; Photographers' Gallery
Foreign Distr: Arts Council of Great Britain (GB)

Profiles British photojournalist Bert Hardy (b.1913), leading staff photographer for the illustrated magazine *Picture Post*.

BETTY HAHN

Series Title: Photoprofiles
30 min. col. video 1983
United States English
Dir.: John Morgan; Rebecca L. Morgan; Paul R. Schranz; Margaretta K. Mitchell **Prod.:** Thomas R. Schiff
Prod. Agency: Images Productions
Distributor (US): Images Productions

Profiles American photographer Betty Hahn (b.1940), known for her experimentation with the gum-bichromate, Vandyke or brown-print, and cyanotype processes. Hahn took up nonsilver printing under the influence of her mentor Henry Holmes Smith. She traces her work from the early stitched gum prints, to the *Lone Ranger* series, to the *Passing Shots* series photographed with a child's Mick-a-matic camera. She also discusses *Cutflowers*, a series of photolithographs, and *Botanical Layouts*, a series of Polaroid prints. Photoprofiles series.

BEYOND THE OCEAN: BENEATH A LEAF
Series Title: Smithsonian Profiles
30 min. col. video 1983
United States English
Prod.: David Feingold
Prod. Agency: Smithsonian Institution Office of Museum Programs; Warner Amex Cable Communications
Distributor (US): Smithsonian Institution Office of Museum Programs
Distributor (US): Britannica Films

Profiles photographer Kjell Sandved, a photographer at the National Museum of Natural History in Washington, D.C., where he documents the work of Smithsonian Institution scientists. Using his camera, he offers compelling observations of the world of nature and insight into the realm of the museum professional. Smithsonian Profiles series, one of three.

BILL BRANDT
Series Title: Master Photographers; No. 2
35 min. col. 16mm 1982
Great Britain English
Prod.: Peter Adam
Prod. Agency: BBC-TV; MM Chanowski Productions bv
Foreign Distr: Arts Council of Great Britain Film Library (GB)

Profile of British photographer Bill Brandt (1904-1983). Brandt is legendary for his refusal to talk about his work, but here he shows some of his favorite pictures, including views of Paris in the 1920s, Great Britain during the Depression, nudes, and several portraits. Master Photographers series, No. 2 of 6.

BILL CHRISTENBERRY'S TRIP TO ALABAMA
Series Title: Miniatures; No. 4
10 min. col. video 1986
United States English
Prod.: Jeffrey Bieber
Exec. Prod.: Joyce Campbell
Prod. Agency: WETA-TV
Distributor (US): WETA-TV

American photographer and sculptor William Christenberry (b.1936) creates visual poems on the "aesthetics of the aging process" in his photographs and sculptures that result from his yearly trips to rural Alabama. He also evokes the tragedy of the South — the Ku Klux Klan and the Civil War — in his more surreal sculptural monuments and photographs. Miniatures series, No. 4 of 4.

BILL PARKER
31 min. b&w video 1979
United States English
Prod. Agency: Video Data Bank
Distributor (US): Video Data Bank

Candida Finkel interviews photography critic Bill Parker, who sees photography as the perfect medium for integrating the real with the ideal and sensation with spirituality. Parker discusses the work of photographers Edward Weston, Murray Riss, Les Krims, Jerry Uelsmann, Minor White, and Robert Frank in terms of Jungian themes and archetypal evidence.

BLACK HAS ALWAYS BEEN BEAUTIFUL
18 min. b&w 16mm; video 1971
United States English
Prod.: Philip Burton, Jr.
Prod. Agency: NET
Distributor (US): University of Illinois Film Center
Distributor (US): Indiana University Audio Visual Center
Distributor (US): Rental-see *Educational Film & Video Locator*
Foreign Distr: Canadian Film Institute (CAN)

James Van Der Zee (1886-1983), self-taught black American photographer, discusses the viewpoints and techniques that made him famous after six decades of obscurity. Van Der Zee, who is noted for his pictures capturing the pride and beauty of black people, shows photographs of Harlem schoolchildren, the Black Yankees, Bill "Bojangles" Robinson, Marcus Garvey, and many others.

BOLTANSKI
Series Title: Magazine des arts
15 min. col. 16mm 1983
France French
Dir.: Alain Fleischer
Prod. Agency: Les Films Constance
Foreign Distr: Les Films Constance (FR)

Profile of French photographer and conceptual artist Christian Boltanski (b.1944), who discusses the creation of his works exhibited at the Centre Georges Pompidou, Paris, in March 1984. Magazine des arts series.

BOLTANSKI
Series Title: Contemporains, Les
26 min. col. 16mm 1984
France French
Dir.: Pierre André Boutang **Prod.:** Jean Michel Meurice

Prod. Agency: Antenne 2
Foreign Distr: Antenne 2 (FR)
Presents the work of French photographer and conceptual artist Christian Boltanski (b.1944). Les Contemporains series.

BRASSAI

49 min. col. 16mm 1971
France French
Dir.: Claude Gallot **Prod.:** Michèle Arnaud
Prod. Agency: INA; ORTF
Foreign Distr: INA (FR)
Profiles French photographer, sculptor, writer, and filmmaker Brassaï (Gyula Halász, 1899-1984), author of *Paris by Night*. Through his reminiscences, he reveals himself as a vagabond who loves to wander.

BRIDGEMEN, THE

24 min. col. video 1986
United States English
Dir.: Paul Turner **Prod.:** Jeff Cretcher; Joel Skidmore
Prod. Agency: Turner Productions
Distributor (US): San Francisco Production Group
A tribute to one of America's first photojournalists, Peter Stackpole (b.1913), and the men who built two of the outstanding constructions of the Depression era, the Golden Gate Bridge (1933-1937) and the San Francisco-Oakland Bay Bridge (opened 1936) in California.

BUSINESS OF PHOTOGRAPHY, THE

Series Title: On Assignment
90 min. col. video 1987
United States English
Dir.: K.C. Lynch **Prod.:** Sally Lynch
Exec. Prod.: Brian D. Ratty
Prod. Agency: Media West Home Video
Distributor (US): Media West Home Video (Video sales only)
Distributor (US): Baker & Taylor Video (Video sales only)
Host photographer Brian D. Ratty examines the business aspects of photography. Covers such subjects as choosing career direction, portfolios, editing, interviews, staff opportunities, stock agencies, major markets, and freelancing. On Assignment series, one of seven.
Reviews: *Shutterbug Magazine*

CAMERA: EARLY PHOTOGRAPHY: CAUGHT IN TIME

Series Title: Camera: Early Photography; No. 8
26 min. col. 16mm 1979
Great Britain English
Dir.: Noel Chanan **Prod.:** Gus Macdonald; Maxine Baker
Prod. Agency: Granada Television International Ltd.
Foreign Distr: Granada Television International Ltd. (GB)
Examines how the camera has recorded important events in history, such as the construction of buildings, Chartist riots in England, and European revolutions. Camera: Early Photography series, No. 8 of 13.

CAMERA: EARLY PHOTOGRAPHY: ENDS OF ART

Series Title: Camera: Early Photography; No. 12
25 min. col. 16mm 1979
Great Britain English
Dir.: David Naden **Prod.:** Gus Macdonald; Maxine Baker
Prod. Agency: Granada Television International Ltd.
Foreign Distr: Granada Television International Ltd. (GB)
Discusses the intimate and complex relationship between art and photography in the nineteenth century. Relates how photographers fought to have their work recognized as art, while artists who denigrated their medium actually painted from photographs, adapting them to produce a desired effect. Camera: Early Photography series, No. 12 of 13.

CAMERA: EARLY PHOTOGRAPHY: EYEWITNESS OF WAR

Series Title: Camera: Early Photography; No. 2
26 min. col. 16mm 1979
Great Britain English
Dir.: John Pett **Prod.:** Gus Macdonald; Maxine Baker
Prod. Agency: Granada Television International Ltd.
Foreign Distr: Granada Television International Ltd. (GB)
Looks at examples of early war photographs, both real and staged, including pictures of the Prussian defeat of France, the Russo-Japanese war, and the record of the British war effort in the Crimea made by British photographer Roger Fenton (1819-1869). Camera: Early Photography series, No. 2 of 13.

CAMERA: EARLY PHOTOGRAPHY: FIRST IMPRESSIONS

Series Title: Camera: Early Photography; No. 1
26 min. col. 16mm 1979

Great Britain English
Dir.: John Pett **Prod.:** Gus Macdonald; Maxine Baker
Prod. Agency: Granada Television International Ltd.
Foreign Distr: Granada Television International Ltd. (GB)

Presents the pioneers of early photography, including British photographer William Henry Fox Talbot (1800-1877) and French photographer Louis Jacques Mandé Daguerre (1787-1851). Examines the development of photography during the Victorian era. Camera: Early Photography series, No. 1 of 13.

CAMERA: EARLY PHOTOGRAPHY: MIRROR IMAGE

Series Title: Camera: Early Photography; No. 4
26 min. col. 16mm 1979
Great Britain English
Dir.: John Pett **Prod.:** Gus Macdonald; Maxine Baker
Prod. Agency: Granada Television International Ltd.
Foreign Distr: Granada Television International Ltd. (GB)

Looks at portrait photographers of the nineteenth century, including British photographer Julia Margaret Cameron (1815-1879) and British writer and mathematician Lewis Carroll (1832-1898) who was also a gifted amateur photographer. Camera: Early Photography series, No. 4 of 13.

CAMERA: EARLY PHOTOGRAPHY: NEWSFLASHES

Series Title: Camera: Early Photography; No. 10
26 min. col. 16mm 1979
Great Britain English
Dir.: Noel Chanan **Prod.:** Gus Macdonald; Maxine Baker
Prod. Agency: Granada Television International Ltd.
Foreign Distr: Granada Television International Ltd. (GB)

Discusses the development of action photography, which progressed as the necessary exposure time shortened. Talks about experiments with different light sources for photography in the dark. These culminated in the birth of the journalist photographer at the end of the nineteenth century. Camera: Early Photography series, No. 10 of 13.

CAMERA: EARLY PHOTOGRAPHY: OUTBACK AND BEYOND

Series Title: Camera: Early Photography; No. 9
26 min. col. 16mm 1979
Great Britain English
Dir.: Martin Smith **Prod.:** Gus Macdonald; Maxine Baker
Prod. Agency: Granada Television International Ltd.
Foreign Distr: Granada Television International Ltd. (GB)

Looks at the photographic records of the massive emigration from Europe to Australia during the nineteenth century. Camera: Early Photography series, No. 9 of 13.

CAMERA: EARLY PHOTOGRAPHY: PHOTOGRAPHIC PLEASURES

Series Title: Camera: Early Photography; No. 5
26 min. col. 16mm 1979
Great Britain English
Dir.: Martin Smith **Prod.:** Gus Macdonald; Maxine Baker
Prod. Agency: Granada Television International Ltd.
Foreign Distr: Granada Television International Ltd. (GB)

Considers the popular appeal of photography during the Victorian era and the birth of the snapshot toward the end of the nineteenth century, when Kodak cameras became widely available. Camera: Early Photography series, No. 5 of 13.

CAMERA: EARLY PHOTOGRAPHY: SOLDIERS OF THE QUEEN

Series Title: Camera: Early Photography; No. 6
26 min. col. 16mm 1979
Great Britain English
Dir.: Noel Chanan **Prod.:** Gus Macdonald; Maxine Baker
Prod. Agency: Granada Television International Ltd.
Foreign Distr: Granada Television International Ltd. (GB)

Presents military photography of the nineteenth century, showing British armies serving throughout the British Empire. The camera was also used to record the life of the native inhabitants in far-flung reaches of the British Empire and the private life of British rulers in India. Camera: Early Photography series, No. 6 of 13.

CAMERA: EARLY PHOTOGRAPHY: THE OTHER HALF

Series Title: Camera: Early Photography; No. 3
26 min. col. 16mm 1979
Great Britain English
Dir.: Martin Smith **Prod.:** Gus Macdonald; Maxine Baker
Prod. Agency: Granada Television International Ltd.
Foreign Distr: Granada Television International Ltd. (GB)

Discusses how the camera was used to record the social extremes of the nineteenth century, such as city slums and madhouses. Camera: Early Photography series, No. 3 of 13.

CAMERA: EARLY PHOTOGRAPHY: TRUTH AND BEAUTY

Series Title: Camera: Early Photography; No. 11
26 min. col. 16mm 1979

Great Britain English
Dir.: John Pett **Prod.:** Gus Macdonald; Maxine Baker
Prod. Agency: Granada Television International Ltd.
Foreign Distr: Granada Television International Ltd. (GB)

Looks at the battle that raged between two types of photographers in the Victorian era: those who, anxious to have photography accepted as an art, "staged" their photographs and used models; and those who believed in the power of photography to show the natural appearance of people and places. Shows examples of works by British photographers Peter Henry Emerson (1856-1936) and Frank Meadow Sutcliffe (1853-1941). Camera: Early Photography series, No. 11 of 13.

CAMERA: EARLY PHOTOGRAPHY: WHEEL OF LIFE — MOVEMENT

Series Title: Camera: Early Photography; No. 13
26 min. col. 16mm 1979
Great Britain English
Dir.: David Naden **Prod.:** Gus Macdonald; Maxine Baker
Prod. Agency: Granada Television International Ltd.
Foreign Distr: Granada Television International Ltd. (GB)

Presents the pioneering photographic experiments in motion of British photographer Eadweard Muybridge (1830-1904) and French physiologist, inventor, and photographer Jules Etienne Marey (1830-1904). Camera: Early Photography series, No. 13 of 13.

CAMERA: EARLY PHOTOGRAPHY: WORLD ON A PLATE

Series Title: Camera: Early Photography; No. 7
26 min. col. 16mm 1979
Great Britain English
Dir.: John Pett **Prod.:** Gus Macdonald; Maxine Baker
Prod. Agency: Granada Television International Ltd.
Foreign Distr: Granada Television International Ltd. (GB)

Looks at travel photography in the Victorian era. Features pictures of the Himalayas by British photographer Samuel Bourne (1834-1912) and photographs of the Far East by Scottish photographer John Thomson (1837-1921). Also shows photographs of Japan, the American West, and the second Antarctic expedition (1910-1912) of British explorer Robert Scott (1868-1912). Camera: Early Photography series, No. 7 of 13.

CARLETON WATKINS

5 min. col. video 1983
United States English
Dir.: William Howze
Prod. Agency: Amon Carter Museum
Distributor (US): Amon Carter Museum

Introduces the life and work of American photographer Carleton E. Watkins (1829-1916). Between 1854 and 1891 he documented Yosemite in central California and other wilderness areas, as well as the industrial and architectural progress of the American West from southern California to Arizona. Produced for the Watkins retrospective held in 1983 at the Amon Carter Museum, Fort Worth, Texas.

CARLOTTA CORPRON: DESIGNER WITH LIGHT

10 min. col. video 1980
United States English
Prod. Agency: Amon Carter Museum
Distributor (US): Amon Carter Museum

Profile of American photographer Carlotta Corpron (b.1901). A photography teacher at North Texas State University, Corpron explains how she uses paper, mirrors, plastic, and other simple devices to modulate light in her photographs.

CARTES POSTALES VIDEO

28 min. col. 16mm; video 1978
France French
Prod. Agency: I.N.A.; FR3; Antenne 2
Foreign Distr: Intermédia (FR)
Foreign Distr: Antenne 2 (FR)

Shows a selection from a collection of 450 postcards of tourist sites around the world. Special effects, animation, and the soundtrack are used to turn each postcard into a painting, a story, an invitation to travel.

CE QUI ME MEUT

26 min. col. 35mm 1989
France French
Dir.: Cédric Klapisch **Prod.:** Adeline Lecallier
Prod. Agency: La Sept; Productions Lazennec; Région Cable
Foreign Distr: Productions Lazennec (FR)
Foreign Distr: La Sept (FR)

A fictional portrait of French physiologist and photographer Jules Etienne Marey (1830-1904), inventor of chronophotography, a precursor of cinematography. Presented in the style of early-twentieth-century archival footage.

CHEVAL DE FER, LE

8 min. col. 16mm; 35mm 1984
Belgium? French (Dutch and English language versions also avail.)
Dir.: Pierre Levie; Gerald Frydman
Foreign Distr: French Ministry of Foreign Affairs (FR)

British photographer Eadweard Muybridge (1830-1904), to settle a wager with American railroad magnate Leland Stanford, created a photographic device that enabled him to establish that a galloping horse, at top speed, had all four hooves off the ground simultaneously. His motion studies and inventions laid the foundation for motion pictures.

CHINE, VUES INTERIEURES

52 min. col. video 1988
France French
Dir.: Christian Hirou; Karl Kugel
Prod. Agency: Rencontres Internationales de la Photographie d'Arles; Amitiés Franco-Chinoises
Foreign Distr: Films Grain de Sable (FR)

Documents meetings of photographers in Peking, Canton, Chengdu, and Yulin, China.

CINDY SHERMAN: AN INTERVIEW

10 min. col. video 1980
United States English
Dir.: Carole Ann Klonarides; Michael Owen **Prod.:** Carole Ann Klonarides; Michael Owen; Cindy Sherman
Prod. Agency: MICA-TV
Distributor (US): MICA-TV
Distributor (US): Electronic Arts Intermix
Distributor (US): American Federation of Arts
Distributor (US): Video Data Bank

Filmmaker Carole Ann Klonarides and American photographer Cindy Sherman (b.1954) set up a mock interview. As they talk, Sherman takes on various disguises drawn from her photographs. At the same time a series of slides is projected in the background and the Muzak film score changes.

CIRCLE OF LIGHT: THE PHOTOGRAPHY OF PAMELA BONE

32 min. col. 16mm; video 1972
Great Britain Nonverbal
Dir.: Anthony Roland **Prod.:** Anthony Roland
Prod. Agency: Anthony Roland Films
Distributor (US): The Roland Collection

Surveys the work of photographer Pamela Bone through an assemblage of her unique glass transparencies. Inspired by nature, Bone's images are mysterious and dreamlike, yet firmly rooted in the real world. Explores her efforts to develop new techniques to realize her inner vision. Composed entirely of photographic slides.
Awards: Cork Film Festival First Prize, 1972

CIVIL WAR PHOTOGRAPHERS: WITNESS TO THE STORM

43 min. col. 16mm; video 1987
United States English
Distributor (US): Lucerne Media

Combines dramatization and present-day documentary footage of sites to examine the work of the photojournalists who covered the American Civil War.

CLIMATE OF NEW YORK, THE

8 min. b&w 16mm; video 1980
United States English
Dir.: Edgar B. Howard **Prod.:** Edgar B. Howard
Prod. Agency: Checkerboard Foundation, Inc.; Tuckernuck Productions
Distributor (US): artsAmerica
Distributor (US): American Federation of Arts

Film and still photographs by Swiss-born American film and photo artist Rudy Burckhardt (b.1914) illustrate poems by American dance critic and writer Edwin Denby (b.1903) written between 1938 and 1944 all on the topic of New York City. Denby recites his poetry.

COLLEGE ET L'ALBUM, LE

Series Title: Des Images et des Sons
15 min. col. 16mm 1981
France French
Dir.: D. Brillaud
Prod. Agency: C.N.D.P.
Foreign Distr: C.N.D.P. (FR)
Foreign Distr: Cinémathèque de l'Enseignement (FR)

Discusses the role of photography in the teaching of art and the importance of framing in defining an image. Des Images et des Sons series.

COLOR FROM LIGHT

11 min. col. 16mm; video 1976
United States English
Prod.: Clark Dugger
Distributor (US): Churchill Films
Distributor (US): University of Illinois Film Center

Distributor (US): Indiana University Audio Visual Center (16mm rental only)
Distributor (US): Rental-see *Educational Film & Video Locator*

Examines the characteristic spectrum of white light. Uses animation to explain the additive and subtractive color principles and their applications in the creation of visual images in television and photography.

COLOURFUL NOTIONS

Series Title: Horizon
50 min. col. video 1986
Great Britain English
Prod.: John Ross
Prod. Agency: BBC-TV
Foreign Distr: BBC Enterprises (GB)

Presents American inventor and industrialist Edwin Land (b.1909), inventor of the Polaroid camera, who discusses his new theory of color vision, which explains visual effects never before understood. Land recreates some of his early experiments to show how black-and-white photographs, illuminated with only red and white light, can appear in full color and how colors remain constant despite changing illumination. He demonstrates his new machine, the "mouse," which, when run over a black-and-white photograph, can predict the object's true colors. Shows that experiments with rhesus monkeys support Land's theory. Horizon series.

CONNECTIONS: 9-COUNTDOWN

Series Title: Connections; No. 9
52 min. col. 16mm; video 1978
United States; Great Britain English
Prod.: James Burke
Prod. Agency: BBC-TV; Time-Life Films
Distributor (US): Ambrose Video Publishing
Distributor (US): Pennsylvania State University Audio Visual Services
Distributor (US): University of Illinois Film Center
Distributor (US): Rental-see *Educational Film & Video Locator*

Begins with the limelight on an Irish mountain and ends with Thomas Alva Edison's motion pictures. Inventor George Eastman's production of celluloid from guncotton, the magic lantern of an Austrian ballistics teacher, and Eadweard Muybridge's journey to California in 1872 to settle a bet by photographs are related occurrences. Connections series, No. 9 of 10.

CONTACTS

Series Title: Contacts
10 programs, 13 min. ea. col. video 1989
France French
Prod. Agency: La Sept; Riff International Production; Centre National de la Photographie
Foreign Distr: Intermédia (FR)
Foreign Distr: Riff International Production (FR)

Series profiling the world's leading photographers. Programs 2 and 3 profile photographers Sebastiao Salgado (b.1944) of Brazil and Robert Doisneau (b.1912) of France. Salgado's photographs are charged with emotional and symbolic content. In 1985, he won the World Press Prize for his photographs of the drought in Ethiopia. Doisneau is known for his images of everyday life in Paris and its environs. Another program in the series profiles American photographer, painter, and filmmaker William Klein (b.1926). Series also includes programs on American photographer Elliott Erwitt, French photographers Raymond Depardon and Henri Cartier-Bresson, Australian photographer Helmut Newton, and Czech photographer Josef Koudelka. Contacts series, thirteen programs.

CONTEMPORARY PHOTOGRAPHERS: AL WEBER

Series Title: Contemporary Photographers
11 min. col. 16mm 1981
United States English
Prod. Agency: Buckman-Burrill Films
Foreign Distr: Edward Patterson Associates Ltd. (GB)

American photographer Al Weber talks about his work. Contemporary Photographers series, one of five.

CONTEMPORARY PHOTOGRAPHERS: DAVID CAVAGNARD

Series Title: Contemporary Photographers
11 min. col. 16mm 1981
United States English
Prod. Agency: Buckman-Burrill Films
Foreign Distr: Edward Patterson Associates Ltd. (GB)

American photographer David Cavagnard talks about his work. Contemporary Photographers series, one of five.

CONTEMPORARY PHOTOGRAPHERS: MORLEY BAER

Series Title: Contemporary Photographers
11 min. col. 16mm 1981
United States English

Prod. Agency: Buckman-Burrill Films
Foreign Distr: Edward Patterson Associates Ltd. (GB)
American photographer Morley Baer (b.1916) talks about his work. Contemporary Photographers, one of five.

CONTEMPORARY PHOTOGRAPHERS: RUTH BERNHARD

Series Title: Contemporary Photographers
11 min. col. 16mm 1981
United States English
Prod. Agency: Buckman-Burrill Films
Foreign Distr: Edward Patterson Associates Ltd. (GB)
American photographer Ruth Bernhard (b.1905) talks about her work. Contemporary Photographers series, one of five.

CONTEMPORARY PHOTOGRAPHERS: STEVE CROUCH

Series Title: Contemporary Photographers
11 min. col. 16mm 1981
United States English
Prod. Agency: Buckman-Burrill Films
Foreign Distr: Edward Patterson Associates Ltd. (GB)
American photographer Steve Crouch talks about his work. Contemporary Photographers series, one of five.

CONTINUOUS DIARY, THE

13 min. col. 16mm; video 1977
Great Britain English
Dir.: Alan Yentob **Prod.:** Leslie Megahey
Prod. Agency: BBC-TV
Foreign Distr: Arts Council of Great Britain (GB)
Profiles British artist Ian Breakwell (b.1943). Focuses on his diary, a day-to-day record juxtaposing handwritten and typed text with drawing, collage, and photography. Breakwell explains the significance of the diary and of the images he has chosen. Uses a collage narrative — of Breakwell taking photographs, writing, and observing — similar to that of the diary itself.

CONVERSATION WITH LOTTE JACOBI, A

Series Title: Woman
30 min. col. video 1977
United States English
Dir.: Will George **Prod.:** Sandra Elkin
Prod. Agency: WNED-TV
Distributor (US): WNED-TV
Distributor (US): International Center of Photography (Archival only)
German-born American photographer Lotte Jacobi (b.1896) discusses her portraits of famous people, including Albert Einstein, Marc Chagall, and Eleanor Roosevelt. Also talks about her photogenics (cameraless photographs). Woman series.

CONVERSATIONS WITH ROY DECARAVA

28 min. col. 16mm; video 1983
United States English
Dir.: Carroll Parrott Blue **Prod.:** Carroll Parrott Blue
Prod. Agency: WHET-TV
Distributor (US): First Run/Icarus Films
Explores the life and work of American photographer Roy DeCarava (b.1919), contributor to "The Family of Man" exhibition (1955) and the first black photographer to receive a Guggenheim fellowship. Interweaves 108 of his black-and-white stills, including his series of jazz portraits featuring Billie Holiday, John Coltrane, and Roy Haynes. DeCarava discusses his past struggles, his efforts to foster young black photographers, and the relationship of his work to the black American experience. Includes appearances by photographer Ansel Adams, photography critic A.D. Coleman, and Mary Schmidt Campbell, art historian and then executive director of The Studio Museum in Harlem, New York City. Narrated by writer Alex Haley.
Reviews: *Landers Film Reviews*, Spring 86; *Library Journal*, Aug 85
Awards: American Film Festival Blue Ribbon, 1984; Black American Cinema Awards Second Prize

CONVERSATIONS WITH WILLARD VAN DYKE

59 min. col.& b&w 16mm; video 1981
United States English
Dir.: Amalie Rothschild **Prod.:** Amalie Rothschild; Austin Lamont
Prod. Agency: Anomaly Films
Distributor (US): Anomaly Films
Distributor (US): Cinema Guild
Distributor (US): Museum of Modern Art Circ. Film Library
American still photographer and filmmaker Willard Van Dyke (1906-1986) reminisces with longtime friends Ralph Steiner and Joris Ivens, discusses his career in photography and his directorship of the Department of Film at the Museum of Modern Art in New York City, and comments on his successes, failures, and regrets. Includes excerpts

from classic documentary films on which Van Dyke worked as cameraman, such as THE RIVER, THE CITY, and VALLEY TOWN, and from programs he produced for several television series, as well as footage from the film THE PHOTOGRAPHER about Edward Weston.
Reviews: *Film Comment*, Apr 86; *Booklist; San Francisco Chronicle; The Baltimore Sun; Variety; Los Angeles Times; Soho News*
Awards: San Francisco International Film Festival Bronze Award, 1981; CINE Golden Eagle, 1981; Sinking Creek Film Celebration Cash Prize; Los Angeles International Film Exposition, 1981; Athens International Film Festival Special Merit Award, 1981

CORNELL CAPA, PARTS I & II

Series Title: Photography: Aesthetics and Practice
60 min. col. video 1975
United States English
Dir.: Donald Woodman **Prod.:** John Keshishoglou
Distributor (US): International Center of Photography (Archival only)

Profiles American photographer Cornell Capa (b.1918), a "documentary, commentary photographer" and founding director of the International Center of Photography in New York City. Capa talks about his adventure-filled life and shows his photographs of political events, personalities, other cultures, and the art world. Photography: Aesthetics and Practice series.

CORPS PERDU

5 min. col. 35mm 1982
France French
Dir.: José Grisel **Prod.:** José Grisel
Prod. Agency: Cinématographe Expansion
Foreign Distr: Cinématographe Expansion (FR)

Presents photographs and postcards of nudes from the beginning of the twentieth century.

CORREO CULTURAL

Series Title: Correo Cultural
26 min. col. 16mm 1979
France French (Spanish and English language versions also avail.)
Dir.: Rodrigo Castano
Prod. Agency: Interaudiovisuel; Realizacion de Communicaciones Visuales S.A.
Foreign Distr: Intermédia (FR)
Foreign Distr: Fédération Européene des industries techniques du cinéma (FR)

Series of interviews with Latin American artists who have chosen to live and work in Europe, often in France. Includes sequences on Colombian photographer Ignacio Gomez Pullido, Venezuelan artist Jésus Rafael Soto, Argentine artist Julio Le Parc, Mexican painters José Clemente Orozco, Cristina Rubalcava, and José Luis Cuevas, Colombian painter Fernando Botero, Colombian artist Luis Caballero, Peruvian artist Alberto Quintanilla, Cuban painter Wilfredo Lam, Argentine sculptors Alicia Penalba and Maria Orenzanz, Cuban sculptor Augustín Cardenas, Chilean painter José Balmes, and Venezuelan painter Asdrubal Colmenares. Correo Cultural series.

COULEUR ET PHOTO

30 min. col. 16mm 1981
France French
Foreign Distr: Intermédia (FR)
Foreign Distr: Association pour la Diffusion de la Pensée Française (FR)

Describes the principles of color reproduction in photography and the applications of color photography in science.

CRYSTALS

6 min. col. 16mm 1983
Canada English
Dir.: Peter Lipskis
Prod. Agency: Peter Lipskis Films
Foreign Distr: Peter Lipskis Films (CAN)
Foreign Distr: Canadian Filmmakers Distribution Centre (CAN)

A cinematic tribute to American photographer and dairy farmer William Bentley (1865-1931), who photographed snowflakes for forty-six winters (1885-1931) and proved that no two of his 5,381 specimens were identical.

DAGUERRE: THE BIRTH OF PHOTOGRAPHY (DAGUERRE OU LA NAISSANCE DE LA PHOTOGRAPHIE)

Series Title: Innovations Techniques de l'Image au Dixneuvième Siècle, Les
29 min. b&w 16mm; 35mm 1964
France English (French, Spanish, and German language versions also avail.)
Dir.: Roger Leenhardt
Prod. Agency: Les Films Roger Leenhardt; Tybanis Film

Distributor (US): FACSEA
Distributor (US): Rental-see *Educational Film & Video Locator*
Foreign Distr: Intermédia (FR)
Foreign Distr: Tybanis Film (FR)
Foreign Distr: French Ministry of Foreign Affairs (FR)
Foreign Distr: Cinémathèque de l'Enseignement (FR)

Uses early-nineteenth-century lithographs, drawings, and rare photographs to trace the birth of photography. Shows sequences of the development by French photographer Louis Jacques Mandé Daguerre (1787-1851) in 1838 of the first commercially feasible process for recording photographs on polished metal plates, based on the earlier work of French photographer Joseph Nicéphore Niépce (1765-1833). Includes examples of other photography pioneers: William Henry Fox Talbot, Hippolyte Bayard, David Octavius Hill, and Frederick Scott Archer, developer of the collodion process. Les Innovations Techniques de l'Image au Dixneuvième Siècle series.

DAGUERREOTYPES

80 min. col. 16mm 1975
France; Federal Republic of Germany French with English subtitles
Dir.: Agnès Varda **Prod.:** Agnès Varda
Prod. Agency: Ciné-Tamaris
Distributor (US): Tom Taplin
Distributor (US): University of Illinois Film Center
Distributor (US): Rental-see *Educational Film & Video Locator*
Foreign Distr: Ciné-Tamaris (FR)

A tribute to French photographer Jacques Louis Mandé Daguerre (1789-1851) by French filmmaker Agnès Varda (b.1928). Shows the residents and shops of the rue Daguerre in Paris. Varda treats her subjects — baker, tailor, butcher, driving instructor, magician, and perfumer — as neighbors and distills from them the atmosphere of middle-class existence on a particular street in Paris.

DARKROOM, THE

Series Title: On Assignment
90 min. col. video 1988
United States English
Dir.: David Healy **Prod.:** David Healy
Exec. Prod.: Brian D. Ratty
Prod. Agency: Media West Home Video
Distributor (US): Media West Home Video (Video sales only)
Distributor (US): Baker & Taylor Video (Video sales only)

Host photographer Brian D. Ratty guides the viewer through basic darkroom techniques. Examines the processes and materials involved in developing both black-and-white and color photographs. Discusses layouts, proofing, enlargements, and special printing methods. Includes advice for building and equipping the darkroom. On Assignment series, one of seven.
Reviews: *Video Choice Magazine*

DAVID BAILEY

Series Title: South Bank Show, The
55 min. col. video 1988
Great Britain English
Dir.: Kim Evans
Prod. Agency: RM Arts; London Weekend Television
Distributor (US): RM Associates/NY

Profiles British photographer and filmmaker David Bailey (b.1938). He began his career in fashion photography in the early 1960s, which he says enabled him to combine interests in photography, women, and money. His varied work includes Salvation Army posters of the 1960s and his current involvement in Greenpeace, the anti-fur movement, and nuclear disarmament. He has also directed his own pop videos and documentaries on American pop artist Andy Warhol, Italian filmmaker Luchino Visconti, and British photographer Cecil Beaton. Bailey now spends most of his time directing commercials because he finds the thirty-second film spots to be the most creative and useful testing grounds for the feature film he hopes to make. The South Bank Show series.

DAVID GOLDBLATT IN BLACK AND WHITE

52 min. col. video 1986
Great Britain English
Dir.: Noel Channing
Prod. Agency: Bernard Joffa Productions in association with Channel Four Television
Distributor (US): Wombat Film & Video

Profiles South African photographer David Goldblatt (b.1934), who for the past thirty years has been producing sensitive black-and-white photographs of the people of South Africa. Goldblatt discusses his personal feelings for his country and says his mission is to express his political views through photographs that are not overtly political.
Reviews: *Library Journal*, 15 Apr 89; *Booklist*, 1 June 89; *Choice*, May 89
Awards: American Film & Video Festival Blue Ribbon, 1989

DAVID GREMP: JESSICA, JULY 15, 1983

6 min. col. video 1983
United States English
Prod. Agency: Museum of Contemporary Art, Chicago
Distributor (US): Museum of Contemporary Art, Chicago

Chicago photographer David Gremp discusses the aesthetic motivation for his use of an 8x10 view camera. He gives a step-by-step demonstration while photographing his daughter.

DAVID HAMILTON

Series Title: Great Photographers
30 min. col. video 198?
Switzerland English?
Foreign Distr: Polivideo S.A. (SWIT)

Profiles photographer David Hamilton (b.1933), known as the "master of eroticism." Examines his books, postcards, posters, and films. Great Photographers series, one of five.

DAVID HARTMAN: THE SHOOTERS

48 min. col. 16mm 1980
United States English
Dir.: Ed Spiegel **Prod.:** Ed Spiegel
Exec. Prod.: David Hartman; Felix Shagin
Prod. Agency: Rodman-Downs, Ltd., Inc., in association with ABC-TV
Distributor (US): Rodman-Downs, Ltd., Inc.

Television host David Hartman, profiles photographers Mary Ellen Mark, David Kennerly, Neil Leifer, Eddie Adams, and Douglas Kirkland, offering a behind-the-scenes look at how they live and work while on assignment.

Awards: CINE Golden Eagle, 1980

DAVID HOCKNEY'S DIARIES

28 min. col. 16mm; video 1970
United States English (German language version also avail.)
Dir.: Michael Blackwood; Christian Blackwood
Prod. Agency: Blackwood Productions Inc.
Distributor (US): Blackwood Productions Inc.
Distributor (US): Rental-see *Educational Film & Video Locator*
Foreign Distr: Arts Council of Great Britain (GB)

Profiles British artist David Hockney (b.1937), who often uses photographic images in his work. Looks at Hockney's photograph album, his personal diary of places and people documented with a Polaroid camera. Images which developed into paintings are juxtaposed with the completed canvases. Shows Hockney at work on just such a transformation: the large double portrait of British fashion designer Ossie Clark and his wife, Celia. Hockney also discusses his etchings for *Grimm's Fairy Tales*.

Reviews: *Landers Film Reviews*, Spring 86

DAVID HOWARD: AN ARTIST'S MEDIUM AND METHODS

28 min. col. video 198?
United States English
Prod. Agency: San Francisco Center for Visual Studies
Distributor (US): San Francisco Center for Visual Studies

Profiles American experimental photographer and painter David Howard (b.1948). Documents a day in his life as he creates videos, photography, and paintings.

Reviews: *San Francisco Chronicle*

DAYBOOKS OF EDWARD WESTON, THE: HOW YOUNG I WAS

Series Title: U.S.A.: Photography; No. 3
28 min. b&w 16mm; video 1965
United States English
Dir.: Robert Katz **Prod.:** Robert Katz
Exec. Prod.: Curt W. Davis
Prod. Agency: KQED-TV for NET
Distributor (US): Indiana University Audio Visual Center
Distributor (US): Pennsylvania State University Audio Visual Services
Distributor (US): Rental-see *Educational Film & Video Locator*

Profiles the life of American photographer Edward Weston (1886-1958) and illustrates his philosophy of photography. Quotes from *The Daybooks of Edward Weston* accompany works from his soft-focus period, his abstract photographs, and work done in Mexico. Two of his sons, his second wife, and one of his former students discuss and evaluate the artist and their relationship to him. U.S.A.: Photography series, No. 3. Part 1 of a 2-part film.

DAYBOOKS OF EDWARD WESTON, THE: THE STRONGEST WAY OF SEEING

Series Title: U.S.A.: Photography; No. 4
28 min. b&w 16mm; video 1965
United States English
Dir.: Robert Katz **Prod.:** Robert Katz
Exec. Prod.: Curt W. Davis

Prod. Agency: KQED-TV for NET
Distributor (US): Indiana University Audio Visual Center
Distributor (US): Pennsylvania State University Audio Visual Services
Distributor (US): Rental-see *Educational Film & Video Locator*

Explains that to American photographer Edward Weston (1886-1958) the strongest way of seeing was to look at the world without preconceived ideas. Works shown reflecting his objectivity and simplicity of expression include his study of Point Lobos, California; his record of California and the western United States; portraits of his cat; samples from his satirical series; and his civil defense series. U.S.A.: Photography series, No. 4. Part 2 of a 2-part film.

DEGAS

Series Title: Art Awareness Collection, The
7 min. col. 16mm 1973
United States English
Dir.: David W. Powell **Prod.:** David W. Powell; Cherrill Anson
Prod. Agency: The National Gallery of Art in cooperation with WETA-TV; Visual Images
Distributor (US): Britannica Films
Distributor (US): The National Gallery of Art
Foreign Distr: Fergus Davidson Associates (GB)

Presents examples of the paintings and graphics of the French artist Edgar Degas (1834-1917) from the collections of The National Gallery of Art, Washington, D.C., together with frames from pioneer motion pictures by British photographer Eadweard Muybridge (1830-1904), to show how Degas's interest in photography influenced his art. The Art Awareness Collection series.

DEGAS DELACROIX DU REEL A L'IMAGINAIRE

13 min. b&w 16mm 1981
France French
Foreign Distr: Intermédia (FR)
Foreign Distr: Association pour la Diffusion de la Pensée Française (FR)

French artists Edgar Degas (1834-1917) and Eugène Delacroix (1798-1863) used chronophotography, a method of time/motion photography invented by French physiologist, inventor, and photographer Jules Etienne Marey (1830-1904), to inspire and elucidate their work.

DERNIER ADIEU, DE JEAN MARC TINGAUD

8 min. col. video 1988
France French
Dir.: Robert Cohen
Prod. Agency: Maison de la Culture de La Rochelle
Foreign Distr: Films Grain de Sable (FR)

Documents the rapport of French photographer Jean Marc Tingaud (b.1947) with his medium.

DIGGING FROM THE AIR

Series Title: Chronicle
50 min. col. video; 16mm 1983
Great Britain English
Prod.: Ben Shepherd
Prod. Agency: BBC Enterprises
Foreign Distr: BBC Enterprises (GB)

Discusses the importance of aerial archaeology in detecting sites invisible on the ground. Shows how aerial photographs can be used to reconstruct a picture of the complete environment of early man. Chronicle series.

DOCUMENTARILY THINE: LEWIS HINE

17 min. col. video 1977
United States English
Prod. Agency: The Brooklyn Museum
Distributor (US): The Brooklyn Museum

Profiles American photographer Lewis W. Hine (1874-1940), whose work focused on social issues such as child labor during the early twentieth century.

DOCUMENTARY URGE, THE: TOM ARNDT

22 min. col. video 1984
United States English
Dir.: Lisa Nebenzahl
Prod. Agency: Minneapolis Institute of Arts
Distributor (US): artsAmerica
Distributor (US): Minneapolis Institute of Arts

American photographer Thomas Arndt (b.1944) uses the camera's descriptive power to present his perceptions of the United States. Arndt is shown photographing at the Minnesota State Fair and readying a 1984 exhibition at the Minneapolis Institute of Arts with Ted Hartwell, curator of photography. After following the presidential campaign of Walter Mondale, the photographer reserved space in the exhibition for photographs made on election night. Discussing his work, Arndt says that he is passionately a

part of the culture that he portrays. Produced in conjunction with the exhibition "Tom Arndt's America" at the Minneapolis Institute of Arts. Narrated by author Garrison Keillor.
Awards: International Film & TV Festival New York Bronze Medal

DOROTHEA LANGE

Series Title: Against the Odds
13 min. col.& b&w video 1988
United States English
Distributor (US): Films for the Humanities

Profiles American photographer Dorothea Lange (1895-1965). Traces the development of her work from society photographs in San Francisco to her documentation of the effects of the Great Depression on American society. Against the Odds series.

DOROTHEA LANGE: THE CLOSER FOR ME

Series Title: U.S.A.: Photography; No. 2
30 min. b&w 16mm; video 1965
United States English
Dir.: Richard Moore; Phillip Greene **Prod.:** Phillip Greene; Robert Katz
Prod. Agency: WNET; KQED-TV
Distributor (US): Museum of Modern Art Circ. Film Library
Distributor (US): Indiana University Audio Visual Center
Distributor (US): Pennsylvania State University Audio Visual Services
Distributor (US): International Center of Photography (Archival only)
Distributor (US): Rental-see *Educational Film & Video Locator*

Concentrates on the work of American photographer Dorothea Lange (1895-1965) with the Farm Security Administration project during the Depression. In describing what was one of the most important documentary projects ever initiated and produced in still photography, Lange tells how she and others were "turned loose in a region to find out what it looked like, what it felt like, and what the human condition was." She reviews these images, her pictures of World War II, and a project on "The New California" that occupied her during the last years of her life. Lange recognizes both the artist's difficulties and desires in achieving thought-provoking photographs on "what is really there." U.S.A.: Photography series; No. 2. Part 2 of a two-part film.

DOROTHEA LANGE: UNDER THE TREES

Series Title: U.S.A.: Photography; No. 1
30 min. b&w 16mm 1965
United States English
Dir.: Richard Moore; Phillip Greene **Prod.:** Phillip Greene; Robert Katz
Prod. Agency: NET; KQED-TV
Distributor (US): Museum of Modern Art Circ. Film Library
Distributor (US): Indiana University Audio Visual Center
Distributor (US): Pennsylvania State University Audio Visual Services
Distributor (US): International Center of Photography (Archival only)
Distributor (US): Rental-see *Educational Film & Video Locator*

Portrait of American photographer Dorothea Lange (1895-1965). Shows her at work on a one-woman, fifty-year retrospective exhibition for the Museum of Modern Art in New York City, while fighting cancer in the last year of her life. As she reviews her work for exhibition, she is self-critical and revealing. During a critique with John Szarkowski, the museum's curator of photography, and in discussions with her family, she goes through a vast number of images that she took all over the world, commenting on the reasons and emotions that moved her to photograph particular scenes. Well known for her work in the Farm Security Administration project, Lange is credited with having photographed one of the most famous images of that collection, *Migrant Mother.* One of the first women photographers, her straightforward and sensitive approach made her one of the foremost documentary photographers of our time. U.S.A.: Photography series, No. 1. Part 1 of a two-part film.

DREI LEBEN DER ILSE BING, DIE

60 min. col. video 198?
Federal Republic of Germany German
Foreign Distr: ZDF (BRD)

Presents the life of German-born American photographer Ilse Bing (b.1899), from her early life in Frankfurt through her career in Paris to her beginning anew in New York.

DU GESTE A L'INFINI: HANS HARTUNG

Series Title: Zig Zag
43 min. col. 16mm 1980
France French
Dir.: Yves Kovacs **Prod.:** Teri Wehn Damisch

Prod. Agency: Antenne 2
Foreign Distr: Antenne 2 (FR)
At his home in Antibes, France, French painter and photographer Hans Hartung (1904-1989) talks about his pictorial approach and photography. Zig Zag series.

DUANE MICHALS (1939-1997)

14 min. col.& b&w 16mm; video 1978
United States English
Dir.: Edgar B. Howard; Theodore R. Haimes
Prod.: Edgar B. Howard; Theodore R. Haimes
Exec. Prod.: Sydie Lansing; Gerrit Lansing
Prod. Agency: Tuckernuck Productions; Suydam Productions
Distributor (US): American Federation of Arts
Distributor (US): artsAmerica
Distributor (US): International Center of Photography (Archival only)
Foreign Distr: Arts Council of Great Britain Film Library (GB)
From the premise that "to photograph reality is to photograph nothing," American commercial photographer and artist Duane Michals (b.1939) has evolved the notion of "sequences": series of still photographs that portray strange and paradoxical events, sexual fantasies, and surreal encounters. Presents a number of these sequences, using a dissolve or fade from one photograph to the next to emphasize their relationship through time and space. Extending this "animated" effect, the filmmakers have worked closely with Michals to produce moving sequences that, like the still photographs, pose the question of the relation between reality and its supposedly faithful representation on film. Michals comments on the nonaccidental appearance of the images, saying they are waiting to be seen. He talks about the effect he attempts to create through his work and his desire to photograph what is essentially unphotographable, that is, to record an emotion rather than an act.
Awards: CINE Golden Eagle, 1979

DYNAMO: SPECIAL PHOTO

Series Title: Dynamo
26 min. col. video 1989
France French
Dir.: Benoît Delepine; Mathias Sanderson
Prod. Agency: La Sept; Telema
Foreign Distr: La Sept (FR)
Cultural magazine in four segments. LES METAMORPHOSES DE FELIX TOURNACHON profiles French photographer Nadar (Gaspard Félix Tournachon, 1820-1910), who photographed all the celebrities of his era. LA MACHINE A REMONTER DANS LE REEL examines the conviction of French poet and critic Charles Baudelaire (1821-1867) that photography was only a process that reproduced reality, not an artistic vocation. Four photographers prove the contrary. SUIS-JE JUIF OU ES-TU ATHEE profiles Patrick Zachmann, press photographer for the agency Magnum Photos and author of the book *Enquête d'identité*. Explores the question of how he defines himself as a Jew. BOLTANSKI profiles French conceptual artist Christian Boltanski (b.1944). His self-reflexive works incorporate banal objects, texts, and photographs that he scrupulously collects and annotates. Dynamo series.

EADWEARD MUYBRIDGE, ZOOPRAXOGRAPHER

60 min. col. 16mm 1975
United States English
Dir.: Thom Andersen **Prod.:** Thom Andersen
Distributor (US): New Yorker Films
Distributor (US): CRM Films
Biographical study of British photographer Eadweard Muybridge (1830-1904), who pioneered the photographic analysis of human and animal motion. Demonstrates his system of cameras activated by sequential trip wires and animates the resultant images through an ingenious combination of photographs and black screen. Muybridge's controversial public and private lives provide the context for a discussion of his contributions to the art and technology of the cinema. Narrated by actor Dean Stockwell.

EARLY PHOTOGRAPHERS, THE

Series Title: History of World Photography, A
52 min. col. 16mm; video 1982
France English (French language version also avail.)
Dir.: Florence Gruère
Prod. Agency: RM Productions; FR3
Distributor (US): Producer Services Group (TV sales)
Surveys techniques and approaches of early photographers who used the medium for scientific and archaeological documentation. Also includes travel photographers and early portraitists. A History of World Photography series, one of four.

EAST MEETS WEST

7 min. col. 16mm 1984
United States English

Dir.: Christine Lombard **Prod.:** Christine Lombard
Distributor (US): Picture Start

A profile of Chinese photographer Tseng Kwong Chi (b.1950), who now resides in New York City. His humorous photographs convey a fresh perspective on the Western world as confronted by a newcomer from a different political and cultural tradition.

Awards: Bucks County Film Festival Special Committee Award, 1985; Chicago Access Corp.'s Festival Silver Can Award, 1985; Baca Festival cash award winner, 1986; Baltimore Film Forum Honorable Mention, 1986; Muddy Film Festival Certificate of Excellence, 1986; Sinking Creek Film Celebration cash award winner, 1986; New School Parsons Film Show Best Documentary, 1986

ECRIVAINS PHOTOGRAPHIES PAR ROBERT DOISNEAU, LES

Series Title: Aujourd'hui en France; No. 201
4 min. col. 16mm; video 1987
France French (English, Spanish, and Arabic language versions also avail.)
Prod. Agency: French Ministry of Foreign Affairs
Distributor (US): Embassy of France
Foreign Distr: French Ministry of Foreign Affairs (FR)
Foreign Distr: ICS (FR)

The residence of French author Honoré de Balzac in Paris is temporarily transformed into a house full of writers thanks to an exhibition of photographs by French photographer Robert Doisneau (b.1912). His portraits convey a sense of humor and tenderness. Aujourd'hui en France series, No. 201.

EDWARD STEICHEN

Series Title: Wisdom
30 min. b&w 16mm 1953
United States English
Dir.: Isaac Kleinerman **Prod.:** Henry Solomon
Prod. Agency: NBC
Distributor (US): Pennsylvania State University Audio Visual Services
Distributor (US): Rental-see *Educational Film & Video Locator*

American photographer Edward Steichen (1879-1973) looks back on his long career as one of the pioneers of modern photography. Describes the circumstances which arose with the advent of World War I that led him to abandon the palette for the camera. Discusses his views on the role of the photographer as an interpreter of man's life and aspirations, as well as his preparations for "The Family of Man" exhibition at the Museum of Modern Art in New York City (1955). Wisdom series.

EDWARD STEICHEN

Series Title: Masters of Photography
28 min. col. video 1964p 1989r
United States English
Prod. Agency: Camera Three Productions
Distributor (US): Camera Three Productions

Profiles American photographer Edward Steichen (1879-1973), who reflects on his life and comments on many of his photographs. Masters of Photography series.

EDWARD WESTON: ARTIST OF THE CAMERA

25 min. col. video 1981
United States English
Dir.: Joan Kuehl; Rudd Simmons **Prod.:** Joan Kuehl
Exec. Prod.: Jain Kelly
Prod. Agency: Pinhole Productions; Joan Kuehl Productions, Inc.
Distributor (US): Pinhole Productions

Looks at the life and work of American photographer Edward Weston (1886-1958) through the eyes of his son, Cole. Stresses Edward Weston's early life and development as an artist and presents examples of his work and darkroom techniques. Includes footage of Weston at work, shot by American documentary filmmaker Willard Van Dyke (1906-1986), and rare family photographs.

EISENSTAEDT — GERMANY

28 min. col. 16mm; video 1982
Federal Republic of Germany; United States English (German language version also avail.)
Dir.: David Hoffman **Prod.:** Karin Davison
Exec. Prod.: Gordon Bowman
Prod. Agency: Varied Directions
Distributor (US): Varied Directions
Distributor (US): Coronet/MTI Film & Video
Distributor (US): American Federation of Arts
Distributor (US): International Center of Photography (Archival only)

Documents the return of German-born American photographer Alfred Eisenstaedt (b.1898) to Germany after an absence of almost fifty years. He recollects his start in photography, philosophizes about his childhood and work, and tells amusing anecdotes about photographing Ernest Hemingway, Winston Churchill, and other notables as

samples of his signature portraits are shown. These works reflect the humanity Eisenstaedt is acclaimed for capturing in his subjects.
Reviews: *EFLA Evaluations*, Sept-Dec 82; *New York Times*, 4 Sept 85; *Film & Video News*
Awards: American Film Festival Blue Ribbon, 1982; Houston International Film Festival Silver Award, 1982; CINE Golden Eagle, 1981; Cork Film Festival Statue of St. Finbarr in the Fine Arts Category, 1981

ELIOT PORTER

Series Title: Photoprofiles
30 min. col.& b&w video 1983
United States English
Dir.: Paul Schranz **Prod.:** Thomas R. Schiff
Distributor (US): Images Productions
Interviews American photographer Eliot Porter (b.1901), emphasizing his delight in nature and his environmental concerns. After receiving degrees in chemical engineering and bacteriology, he taught at Harvard Medical School for ten years. His meeting with Alfred Stieglitz in 1934, who gave him a one-man show in 1938-1939, led Porter to give up medicine and take up photography full-time. He discusses his preference for color film, darkroom equipment, and techniques using the dye-transfer process for his negatives to ensure permanence of color. Photoprofiles series.

ELIOT PORTER'S WORLD

29 min. col. 16mm; video 1987
United States English
Dir.: Amertat Cohn **Prod.:** Amertat Cohn
Prod. Agency: CNI Pictures, Inc.
Distributor (US): CNI Pictures, Inc.
Distributor (US): Direct Cinema Ltd.
Traces the life and work of American photographer Eliot Porter (b.1901) from his childhood on Great Spruce Head Island in Maine to his most recent trips to Antarctica and China. Compares his landscape photographs with the actual locations in Maine and New Mexico. Porter discusses how American photographer Alfred Stieglitz (1864-1946) encouraged him in his career. There are other interviews with Porter's wife, Aline, and with Eleanor Caponigro, his book designer. Porter talks about the creation of his personal aesthetic and his preference for color photography.
Reviews: *Booklist*, 1 Apr 89

ENA'S ADVENTURES: PART ONE

31 min. col. video 1986
United States English
Dir.: Lee Eiferman **Prod.:** Kathy High
Prod. Agency: Mass Productions
Distributor (US): Mass Productions
Tells the story of Ena, an itinerant photographer hired by the Works Progress Adminstration (WPA) in the 1930s to photograph the people of the southwestern United States. As she photographs three subjects in the back of a radio repair shop, she sees images of their past and fantasies through the camera. Later she discovers that the government wants to apprehend her for violating the terms of her WPA agreement. The style of presentation is based on radio serials from the 1930s.

ENFANCE D'UN PHOTOGRAPHE: LARTIGUE

23 min. col. video 1975
France French
Dir.: Georges Rouveyre; G. Bardawil
Prod. Agency: INRP
Foreign Distr: INRP (FR)
French photographer Jacques-Henri Lartigue (1894-1986) comments on a selection of his own photographs taken during his childhood and early adult years. Evokes the aura of the beginning years of the twentieth century.

ENTREE LIBRE

? min. col. 16mm 1984
France French
Dir.: Philippe Kimmerling **Prod.:** Pierre Charpentier; Jean Louis Cros
Foreign Distr: FR3 (FR)
French photographer Robert Doisneau (b.1912) comments on various topics.

EPOQUE ETAIT DROLE, L'

10 min. col. 16mm; video 1982
France French
Dir.: Thierry Marzloff **Prod.:** Thierry Marzloff
Prod. Agency: Cinémage
Foreign Distr: Cinémage (FR)
Foreign Distr: Thierry Marzloff (FR)
Visits a caretaker's lodge at the Place Vendôme in Paris, where Rose-Marie, a famous cover girl from the 1950s, lives and works. She recalls her past without bitterness as she looks at fashion photographs from the *Vogue* magazine exhibition at the Jacquemart-André Museum.

ERNEST LIVERNOIS, PHOTOGRAPHE

53 min. col.& b&w 16mm 1988
Canada French
Dir.: Arthur Lamothe **Prod.:** Pierre Lamy; François Brault
Prod. Agency: Les Films François Brault Inc.
Foreign Distr: Cinéma Libre Inc. (CAN)

Dramatizes an evening in 1898 when Canadian photographer Jules-Ernest Livernois (1851-1933) wonders if his son will succeed him in his photographic studio in Quebec as he had succeeded his father, Jules-Isaï. A childhood friend visits him and they tour the gallery, studio, and laboratory. When the Livernois studio, founded in 1854, closed in 1950, it left behind over three hundred thousand photographs, from black and white to color, from daguerreotypes to pictures using the most modern production techniques.

ESTELLE JUSSIM

28 min. b&w video 1980
United States English
Prod. Agency: Video Data Bank
Distributor (US): Video Data Bank

Amy Amdur interviews communications theorist Estelle Jussim, who discusses her interest in still photography, her disillusionment with semiotics, and her ambition to create new paths for the history of photography.

EUGENE ATGET

10 min. b&w 16mm 1963
United States Nonverbal
Dir.: Harold Becker **Prod.:** Harold Becker
Prod. Agency: Harold Becker Productions
Distributor (US): Biograph Entertainment, Ltd.
Distributor (US): Rental-see *Educational Film & Video Locator*

Paris at the turn of the century, seen in the works of French photographer Eugène Atget (1857-1927). Rare photographs of streets, cafés, shop windows, gardens with statues, chestnuts in bloom, and a barrel organ provide a lyric and nostalgic view of a bygone Paris. Accompanied by the music of Erik Satie.

EUGENE ATGET: PHOTOGRAPHER

50 min. col. 16mm; video 1982
Great Britain English
Dir.: Peter Wyeth **Prod.:** Peter Wyeth
Prod. Agency: Arts Council of Great Britain
Foreign Distr: Arts Council of Great Britain (GB)

A dramatized profile of French photographer Eugène Atget (1857-1927), whose work was only truly appreciated after his death. Among those who bought his work were French painters Maurice de Vlaminck, Georges Braque, and Maurice Utrillo. Man Ray and Berenice Abbott recognized that Atget's photographs were works of art. Shows numerous examples of Atget's photographs and follows the development of his themes around Paris.

EVERYTHING IS PHOTOGRAPH: PROFILE OF ANDRE KERTESZ

Series Title: Masters of Photography
30 min. col. video 1978p 1989r
United States English
Dir.: John Musilli **Prod.:** John Musilli
Prod. Agency: Camera Three Productions
Distributor (US): Camera Three Productions

Profiles Hungarian-born American photographer André Kertész (1894-1985) in the later years of his life. Follows him on a shoot with a model in Washington Square Park near his home in New York City. Shows many of his photographs from early to late works. Kertész discusses his approach to photography, which stresses the subject rather than formal concerns, and relates the importance of shooting decisively and selectively. The photographer also recalls the pivotal years he spent in Paris in the 1920s and 1930s and his impressions of the United States since his arrival in 1936. Masters of Photography series.

EVON STREETMAN

32 min. b&w video 1980
United States English
Prod. Agency: Video Data Bank
Distributor (US): Video Data Bank

Joyce Neimanas interviews American photographer Evon Streetman, who discusses her working methods and her involvement with teaching at the University of Florida at Gainesville.

EXPOSITION A NANCY 1909

? min. col. 16mm 1980
France French
Prod. Agency: FR3
Foreign Distr: FR3 (FR)

The first international photography exposition took place from June to November 1909 in Nancy, France. It brought together eighteen hundred exhibitors in the Parc Sainte-Marie. Today, only an Alsatian house, a music kiosk, and a hot spring remain.

EXTRAITS DU JOURNAL DE JACQUES-HENRI LARTIGUE

Series Title: Italiques
56 min. col. 16mm 1974
France French
Dir.: Claude Ventura **Prod.:** M. Gilbert
Prod. Agency: ORTF
Foreign Distr: INA (FR)
Foreign Distr: Intermédia (FR)

Profiles French photographer Jacques-Henri Lartigue (1894-1986), who from the beginning of the twentieth century patiently amassed thousands of photographs and countless pages in his personal diary. This enormous memoir forms a record of the places, people, and epochs of Lartigue's lifetime. Italiques series.

EYE OF CARTIER-BRESSON, THE

Series Title: Monitor
12 min. b&w 16mm 1960
Great Britain English
Prod. Agency: BBC-TV
Foreign Distr: BBC Enterprises (GB) (Sales only)

Shows a series of still and moving images as French photographer Henri Cartier-Bresson (b.1908) discusses his work. Shot in and around Paris. Monitor series.

FACE TO FACE

30 min. col. 16mm 198?
Australia English
Dir.: Stephen Burstow **Prod.:** Stephen Burstow
Foreign Distr: Australian Broadcasting Commission (AUSTRAL)

Presents the work of three Australian photographers who specialize in images of people and explain why they do so. Gerrit Fokkema takes corporate portraits in boardrooms and offices and photos of friends and family shot in a laundromat. Anne Zahalka makes a portrait with bowls of artificial fruit to suggest the impact of European painting traditions on Australians. John Williams searches his own past to find the meaning of his family snapshots.

FAMILY OF MAN, THE

Series Title: Adventure
30 min. b&w 16mm; 35mm 1956
United States English
Dir.: Joseph Scibetta **Prod.:** Robert Northshield
Prod. Agency: CBS
Distributor (US): Museum of Modern Art Dept. of Film (Archival only)

Documents the photographic exhibition "The Family of Man," organized by American photographer Edward Steichen (1879-1973) at the Museum of Modern Art, New York City, in 1955. Includes interviews with Steichen, American poet Carl Sandburg, former first lady Eleanor Roosevelt, and anthropologist Harry Schapiro. Adventure series.

FAMILY OF MAN, THE

21 min. b&w 16mm 1955
United States English
Dir.: Robert Squier **Prod.:** Robert Squier; Philip Gelb; Stanton Catlin
Distributor (US): Museum of Modern Art Dept. of Film (Archival only)

Documents a speech by American photographer Edward Steichen (1879-1973) introducing "The Family of Man," an exhibition of photographs presented at the Minneapolis Institute of Arts after its initial installation at the Museum of Modern Art in New York City in 1955. Presents a sampling of the photographs as Steichen discusses why he chose them for the exhibition. Toward the end of the speech, Steichen argues against the use of the hydrogen bomb.

FASHION IN THE MAKING

10 min. col. 16mm 1979
United States English
Distributor (US): Pennsylvania State University Audio Visual Services
Distributor (US): Rental-see *Educational Film & Video Locator*

Shows English fashion photographer Norman Parkinson (1913–1990) at work on location in the Bahamas as he photographs models for *Gentlemen's Quarterly*. (*Penn State*)

FAY GODWIN

Series Title: South Bank Show, The
55 min. col. video 1986
Great Britain English
Dir.: Hilary Chadwick
Prod. Agency: RM Arts; London Weekend Television
Distributor (US): RM Associates/NY

Profiles British landscape photographer Fay Godwin (b.1931). Discusses her work in relation to other landscape artists, showing how her camera has taken the place of the painter's brush. Hosted by Melvyn Bragg. The South Bank Show series.

FILM SUR HANS BELLMER

35 min. col. 16mm 1972
France French
Dir.: Catherine Binet
Prod. Agency: Les Films Jean-Noël Delamarre
Foreign Distr: Intermédia (FR)
Foreign Distr: Les Films Jean-Noël Delamarre (FR)

Considers two aspects of the erotic art of German painter, photographer, sculptor, and graphic artist Hans Bellmer (1902-1975). The first part, THE GAMES OF THE DOLL, shows photographs, taken by the artist in the house and garden of his youth, of a doll he designed, whose function is not procreation but "the vertiginous awakening of passion, extended to the invention of desires." On the sound track, Unica Zürn (1916-1970), his companion for seventeen years, until she committed suicide, interprets an old German ballad, while a twelve-year-old girl delivers a few stanzas from a prose poem by Paul Eluard, *The Games of the Doll*. The second part, THE DECREPIT AND THE WRINKLED, is derived from Bellmer's book, *The Small Anatomy of the Physical Unconscious or Anatomy of the Image* (1957), an elaborate text on the thoughts, aims, and desires of the artist, as expressed in his paintings, sculptures, and engravings. (*Montreal Film Festival*)

FILMING MUYBRIDGE

25 min. b&w 16mm 1980
France Nonverbal
Dir.: Jean-Louis Gonnet **Prod.:** Jean-Louis Gonnet
Prod. Agency: Institut d'Arts Visuels d'Orléans
Foreign Distr: Intermédia (FR)
Foreign Distr: Institut d'Arts Visuels d'Orléans (FR)

Brings to life the motion-study images taken by British photographer Eadweard Muybridge (1830-1904).

FIRE IN THE EAST: A PORTRAIT OF ROBERT FRANK

28 min. col. video 1986
United States English
Dir.: Philip Brookman; Amy Brookman **Prod.:** Anne Wilkes Tucker; Paul Yeager
Exec. Prod.: Robert P. Roman (Museum of Fine Arts, Houston); Miriam F. Korshak (KUHT-TV)
Prod. Agency: Museum of Fine Arts, Houston; KUHT-TV
Distributor (US): American Federation of Arts
Distributor (US): KUHT-TV
Distributor (US): Museum of Fine Arts, Houston

Presents the life and work of Swiss-born American photographer Robert Frank (b.1924), who now lives in Nova Scotia, Canada. Includes clips from his film and video work. Interviews with friends and colleagues include photographers Louis Faurer, Elliott Erwitt, Sid Kaplan, Duane Michals, and Ed Grazda; filmmakers Emile de Antonio, Walter Gutman, and Jonas Mekas; artists Miles Forst and June Leaf; poet Allen Ginsberg; and writer Rudy Wurlitzer.
Awards: Educational Film Library Bronze Apple

FIXED IN TIME: A VICTORIAN ALBUM

19 min. col. 16mm; video 1980
Canada English
Dir.: Shelagh Mackenzie **Prod.:** Barry Cowling
Exec. Prod.: Rex Tasker
Prod. Agency: National Film Board of Canada
Foreign Distr: National Film Board of Canada (CAN)

Large glass plate photographs taken by Oliver Massey Hill (1869-1905), manager of the Notman Gallery Studio in Halifax, Nova Scotia, provide a remarkable visual record of the nineteenth-century British garrison town. His candid and often humorous musings about what he saw through his lens, and authentic music of the period, combine to breathe life into the images of this era.
Reviews: *A Newsletter Called FRED...*, Nov-Dec 85

FLAGRANTS DELITS

29 min. b&w 16mm; video 1967
France French
Dir.: Robert Delpire
Prod. Agency: Delphine Editions et Productions
Foreign Distr: Cinéculture (FR)

Presents the work of French photographer Henri Cartier-Bresson (b.1908). Shows over six hundred examples of the photographer's finest work.

FOCUS ON PHOTOGRAPHY

23 min. col. video 1987
United States English
Dir.: Frances C. Stoddard **Prod.:** Frances C. Stoddard
Prod. Agency: International Center of Photography
Distributor (US): Encyclopaedia Britannica Educational Corp.

Traces the development of photography in the nineteenth and twentieth centuries, examining processes, aesthetic concepts, and its changing uses.

FREEMAN IN TIME

6 min. col. 16mm 1979
Canada English **Prod.:** David Latchman
Foreign Distr: David Latchman (CAN)
Foreign Distr: National Film Board of Canada (CAN)

A portrait of Canadian photographer Freeman Patterson (b.1937). Shows a selection of his photographs and sequences of Patterson choosing subjects, photographing, developing film, and explaining his work to a youngster at one of his exhibitions.

FRENCH MUSEUM OF PHOTOGRAPHY, THE

7 min. col. 16mm; video 1981p 1988r
France English (French, Arabic, Portuguese, and Spanish language versions also avail.)
Distributor (US): FACSEA
Foreign Distr: Intermédia (FR)
Foreign Distr: Association pour la Diffusion de la Pensée Française (FR)

Shows the Musée Français de la Photographie in Bièvres, France, founded in honor of French photographer Nadar (Gaspard Félix Tournachon, 1820-1910) and dedicated to the technical and artistic history of photography.

FRIDA KAHLO AND TINA MODOTTI

29 min. col. 16mm; video 1982
Great Britain English
Dir.: Laura Mulvey; Peter Wollen
Prod. Agency: Modelmark Ltd.; Arts Council of Great Britain
Distributor (US): American Federation of Arts
Distributor (US): Museum of Modern Art Circ. Film Library
Foreign Distr: Arts Council of Great Britain Film Library (GB)

Portrays the work of two women artists, Frida Kahlo (1907-1954), painter and wife of artist Diego Rivera, and Tina Modotti (1896-1942), photographer and protégée of photographer Edward Weston, and their struggle to achieve recognition and identity in the male-dominated culture of postrevolutionary Mexico. Analyzes their work and explains that their paths crossed in a mural by Rivera. Shows how each woman defined herself differently in the face of the necessities and accidents of history and biography and in relationship to her own body. Includes rare home movie footage of Kahlo and Rivera shot by photographer Nickolas Muray and a segment of a silent film starring Tina Modotti.

FROM THE LOG OF JUST LOVE

27 min. col. 16mm 1976
United States English
Prod. Agency: Brooks Institute of Photography
Distributor (US): Brooks Institute of Photography (Archival only)

Demonstrates the proper physical training and mental attitude necessary in underwater photography. Shows the different uses and types of underwater photography.

FRUEHE PHOTOGRAPHISCHE BILDNISSE IM KUPFERSTICH-KABINETT

Series Title: Gems of Dresden Collections
15 min. col. video 1989?
German Democratic Republic German
Foreign Distr: Television of the German Democratic Republic (DDR)

Presents a group of photographic portraits taken between 1840 and 1865 by several pioneer photographers. Gems of Dresden Collections series.

GEORGE EASTMAN: ONE MAN'S VISION, IMAGE FOR ALL

Series Title: American Life Style
29 min. col. 16mm 1981
United States English
Dir.: Ann Zane Shanks **Prod.:** Ann Zane Shanks
Prod. Agency: Comco Productions
Distributor (US): Show Corp.

Documents the life of George Eastman (1854-1932), American industrialist, inventor, and philanthropist, who introduced the Kodak camera in 1888, founded the Eastman Kodak Company, and invented various film processes. American Life Style series.
Awards: CINE Golden Eagle, 1981

GEORGE HENDRIK BREITNER

37 min. col. 16mm 1975
Netherlands English (French, German, and Spanish language versions also avail.)
Dir.: Rob Mariouw Smit
Distributor (US): Consulate-General of the Netherlands
Foreign Distr: Netherlands Govt. Information Service (NETH)

Examines the work of Dutch painter, graphic artist, and photographer George Hendrik Breitner (1857-1923), best known as a portraitist of nineteenth-century Amsterdam, where he lived and worked from 1886 until 1914.

GEORGE TICE

Series Title: Photoprofiles
30 min. col. video 1983
United States English
Dir.: John Morgan; Rebecca L. Morgan; Paul R. Schranz; Margaretta K. Mitchell **Prod.:** Thomas R. Schiff
Prod. Agency: Images Productions
Distributor (US): Images Productions

Profiles American photographer George Tice (b.1938). Born in New Jersey, Tice's work explores the objects and locations within miles of his birthplace. Tice discusses growing up in New Jersey, his early work in commercial photography, and his recent preoccupation with President Abraham Lincoln. Photoprofiles series.

GEORGES ROUSSE

10 min. col. 35mm 1985
France Nonverbal
Dir.: Christophe Loizillón
Prod. Agency: Les Films du Lagon Bleu
Foreign Distr: Intermédia (FR)
Foreign Distr: Les Films du Lagon Bleu (FR)

Documents the creative process of French painter and photographer Georges Rousse, who paints optical illusion murals in abandoned buildings slated for demolition, then photographs the paintings. The photos are the only record of the work.

GEORGES ROUSSE

Series Title: Aujourd'hui en France; No. 158
4 min. col. 16mm; video 1985
France French (English, Spanish, and Arabic language versions also avail.)
Prod. Agency: French Ministry of Foreign Affairs
Distributor (US): Embassy of France
Foreign Distr: French Ministry of Foreign Affairs (FR)
Foreign Distr: ICS (FR)

French painter and photographer Georges Rousse works in abandoned building lots in Paris, Berlin, Sydney, and New York. He is shown in Paris on the construction site for the future Opéra de la Bastille and among the warehouses at Bercy, creating paintings which will never be seen except in his photographs of them. Aujourd'hui en France series, No. 158.

GIANPAOLO BARBIERI

Series Title: Great Photographers
30 min. col. video 198?
Switzerland English?
Foreign Distr: Polivideo S.A. (SWIT)

Profiles Italian fashion photographer Gianpaolo Barbieri, whose photographs document the past twenty years of exclusive fashion. Explores his background and his experience with the theater and with film director Luchino Visconti. Reveals some of Barbieri's techniques. Great Photographers series, one of five.

GISELE FREUND: 1982

Series Title: Aujourd'hui en France; No. 71
5 min. col. 16mm; video 1982
France French (English, Spanish, and Arabic language versions also avail.)
Dir.: Elizabeth Lennard
Prod. Agency: French Ministry of Foreign Affairs
Distributor (US): Embassy of France
Foreign Distr: French Ministry of Foreign Affairs (FR)
Foreign Distr: ICS (FR)

Profiles German-born French photographer and journalist Gisèle Freund (b.1912), who fled Nazi Germany in 1933 to live in France. She is best known for her pioneering work in color photography and her sensitive portraits of the leading literary and artistic figures of the twentieth century, among them writers André Malraux, Jean-Paul Sartre, André Gide, Colette, and Virginia Woolf and artists Marcel Duchamp and Henri Matisse. From 1948 to 1954 Freund was associated with the photographers' collaborative, Magnum Photos. She was chosen by French president François Mitterrand to take his official portrait. Filmed in her apartment, Freund discusses her philosophy of life and art and tells anecdotes about her portrait sessions. Aujourd'hui en France series, No. 71.

GLOIRE A FELIX TOURNACHON

11 min. b&w 16mm 1966
France English (French, Spanish, German, Portuguese, Arabic, and Russian language versions also avail.)
Dir.: André Martin; Michel Boschet
Prod. Agency: Argos Films
Foreign Distr: Intermédia (FR)
Foreign Distr: French Ministry of Foreign Affairs (FR)

Profiles French photographer Nadar (Gaspard Félix Tournachon, 1820-1910). Beginning as a journalist and lithographer, he switched to photography in the mid-1850s. He is particularly known for the portraits and views of Paris he photographed during the 1860s and 1870s.

GOING WHERE I'VE NEVER BEEN: PHOTOGRAPHY OF DIANE ARBUS

Series Title: Masters of Photography
30 min. col.& b&w video 1972p 1989r
United States English
Dir.: John Musilli **Prod.:** John Musilli
Prod. Agency: Camera Three Productions
Distributor (US): Camera Three Productions

Discusses the life, career, and aesthetic of American photographer Diane Arbus (1923-1971) with her friends, family, and colleagues. Includes interviews with her daughter Doon Arbus; photographer and teacher Lisette Model; artist and designer Marvin Israel; and John Szarkowski, director of the photography department at the Museum of Modern Art, New York. Mary Costello, a close friend of the artist, reads commentary on Diane Arbus's photographs excerpted from lectures given by the artist shortly before her death. Masters of Photography series.

GORDON PARKS: MOMENTS WITHOUT PROPER NAMES

60 min. col. 16mm 1988
United States English
Distributor (US): Past America Inc.

Examines the life of American photographer, writer, director, and composer Gordon Parks (b.1912). Photographer for *Life* magazine from 1948 to 1968, he directed a major film, THE LEARNING TREE, in 1968. Shows photographs by Parks, uses his poetry as the narrative, and features his original score for the film.

GRACE ROBERTSON

Series Title: Five Women Photographers; No. 1
25 min. col. video 1986
Great Britain English
Dir.: Judy Marle **Prod.:** Ruth Pantoleon
Prod. Agency: Broadside Productions; Channel Four Television
Foreign Distr: Albany Video (GB)

Profiles British photojournalist Grace Robertson, the first woman photographer to join the staff of *Picture Post* magazine. Features an interview with Robertson. Five Women Photographers, No. 1 of 5.

GRAND REPORTER

Series Title: Grand Reporter
4 programs, 52 min. ea. col. video 1989
France; Federal Republic of Germany; Switzerland French
Dir.: Pierre Zucca; Pierre Brard
Prod. Agency: La Sept; FR3; Radio Télévision Suisse Romande; Radio Diffusion Télévision Sud-Ouest; INA; IMA; SSR
Foreign Distr: La Sept (FR)

Four-part series that uses photographs and films to explore the motivations, objectives, methods, and disillusionment of the photographers and reporters who provide the press and television with images of the world. Part 1, LA FORCE DU TEMOIGNAGE, examines the power of the eyewitness account and features the work of Albert Londres, Michel Laurent, Erich Lessing, and Marc Riboud. Part 2, LE RISQUE DE L'ENQUETE, shows the work of some photographers who documented the misery and desperation of the poor: Lewis W. Hine, Dorothea Lange, Walker Evans and James Agee, and Jean-Luc Porquet. Part 3, LA PASSION DE LA DECOUVERTE, asks whether reporters should be outside observers or become involved in the society they cover and presents the work of André Sauvage, Albert Londres, Jean Gaumy, and Gérard Géry. Part 4, LE POUVOIR DE L'INVESTIGATION, analyzes the power and influence of reporting and features prize-winning reporter Frédéric Laffont. Grand Reporter series.

GREAT DAYS IN THE ROCKIES

11 min. b&w 16mm; 35mm; video 1983
Canada English
Dir.: Arvi Liimatainen **Prod.:** Jerry Krepakevich; Anne Wheeler
Exec. Prod.: Tom Radford
Prod. Agency: National Film Board of Canada
Foreign Distr: National Film Board of Canada (CAN)

In 1903, Canadian photographer Byron Harmon (1876-1942) began to photograph Banff, Alberta and the Canadian Rockies: the mountains, the Native Americans, mountain climbers, and European aristocrats in search of new horizons. The film is made entirely of stills and motion pictures shot by Harmon. There are 6,500 negatives of his work at the Peter Whyte Archive in Banff.
Awards: Edmonton Film Festival

GUY LE QUERREC, PHOTOGRAPHE

Series Title: Aujourd'hui en France; No. 91
6 min. col. 16mm; video 1982
France French (English, Spanish, and Arabic language versions also avail.)
Prod. Agency: French Ministry of Foreign Affairs
Distributor (US): Embassy of France
Foreign Distr: French Ministry of Foreign Affairs (FR)

Foreign Distr: ICS (FR)

Guy Le Querrec, a French photographer with the agency Magnum Photos, is a devoted jazz fan. Shows him in his native Britanny, running a workshop combining photography and jazz music. Aujourd'hui en France series, No. 91.

HAL FISCHER READS *AMERICAN PHOTOGRAPHER*: T & A MADE EASY

30 min. col. video 1985
United States English
Prod. Agency: Paper Tiger Television
Distributor (US): Paper Tiger Television

American writer and critic Hal Fischer (b.1950) critiques the magazine *American Photographer*.

HANS BELLMER

Series Title: Variances
20 min. b&w 16mm 1972
France French
Dir.: Alain Jouffroy; Michel Pamart **Prod.:** Michèle Arnaud
Prod. Agency: INA
Foreign Distr: INA (FR)

Profiles German painter, photographer, sculptor, and graphic artist Hans Bellmer (1902-1975), who lived for a number of years in France. Variances series.

HANS NAMUTH, INSTANTS DONNES

26 min. col. video 1988
France French
Dir.: Teri Wehn Damish; Jean-Michel Vecchiet
Prod. Agency: Rencontres Internationales de la Photographie d'Arles; Amitiés Franco-Chinoises
Foreign Distr: Films Grain de Sable (FR)

Looks at the work of German-born American photographer Hans Namuth (b.1915). He photographed the Republican struggle during the Spanish Civil War before enlisting in the French Foreign Legion and the United States Army. He has recorded significant moments in political history, and important figures in the world of art have sat for him, including Jackson Pollock, Willem de Kooning, Constantin Brancusi, Josef Albers, Henri Matisse, and Andy Warhol. Namuth's film POLLOCK documents the unique creative process of painter Jackson Pollock.

HANS STAUB, PHOTOJOURNALIST (HANS STAUB, FOTOREPORTER)

60 min. b&w 16mm 1978
Switzerland German with English subtitles
Dir.: Richard Dindo
Foreign Distr: Pro Helvetia (SWIT)

Profile of Swiss photographer Hans Staub (b.1894), who chronicled the 1930s and World War II years in the Swiss magazine *Zürcher Illustrierte*. Interviews the magazine's editor, Arnold Kübler, who discusses his collaboration with Staub and the historical importance of photojournalism. Staub talks about his photographs and his role as a photojournalist.

HARRY CALLAHAN: A BROADCAST EXHIBITION

24 min. col. 16mm; video 1985
United States English
Dir.: Andrew Silver; Yong-Hee Silver **Prod.:** Andrew Silver; Yong-Hee Silver
Prod. Agency: Silver Productions
Distributor (US): Silver Productions

American photographer Harry Callahan (b.1912) discusses his life and work. Shows photographs by the artist.
Awards: American Film & Video Festival Blue Ribbon, 1986

HARRY CALLAHAN: A NEED TO SEE AND EXPRESS

Series Title: Masters on Photography; No. 1
23 min. col. video 1982
United States English
Prod. Agency: Silver Productions
Distributor (US): Video Gallery
Distributor (US): Guidance Associates

Reveals the diverse photographic interests of American photographer Harry Callahan (b.1912) by featuring more than three hundred of his color and black-and-white photographs and rare footage from his films. Interspersed with an interview of Callahan, which offers biographical clues to the photographer's work. Masters on Photography series, No. 1.
Reviews: *EFLA Evaluations*, Fall-Winter 83/84; *Sightlines*, Summer 83
Awards: American Film Festival Blue Ribbon, 1983

HARRY CALLAHAN: ELEANOR AND BARBARA

18 min. col. 16mm; video 1983
United States English
Dir.: Edgar B. Howard; Seth Schneidman **Prod.:** Edgar B. Howard

Prod. Agency: Checkerboard Foundation, Inc.
Distributor (US): American Federation of Arts
Distributor (US): artsAmerica
Distributor (US): International Center of Photography (Archival only)

Photographs by American photographer Harry Callahan (b.1912) of his wife and daughter, taken over a period of approximately twenty-five years, provide an important insight into his creativity. The photographs of his wife and model, Eleanor, comprise a major study of the nude. Includes interviews with Harry, Eleanor, and Barbara Callahan.

Reviews: *A Newsletter Called FRED...*, Vol 13, No 6; *EFLA Evaluations*, Summer-Fall 85; *Sightlines*, Summer-Fall 85; *New York Times*, 4 Sept 85
Awards: American Film Festival Blue Ribbon, 1985

HARTUNG

12 min. col. 16mm 1982
France French
Dir.: Pierre Devert
Prod. Agency: Centre Georges Pompidou; Cineteve
Foreign Distr: Cinéculture (FR)

Looks at the photography of French painter and printmaker Hans Hartung (1904-1989).

HEALTH HAZARDS IN PHOTOGRAPHY

Series Title: Health Hazards in Art
9 min. col. video 1984
United States English
Prod.: Rachel Stevenson
Prod. Agency: International Film Bureau
Distributor (US): International Film Bureau

Examines health hazards in photographic processing and details darkroom safety practices. Warns against the ingestion and absorption of chemicals and shows the importance of rubber aprons and gloves, goggles, and respirators. Reviews safety devices which should be near the work area. Health Hazards in Art series, one of five.

Reviews: *School Arts*, Dec 84; *School Arts*, May 86; *The Book Report*, Mar-Apr 85; *Media Review*, May 85

HELEN MUSPRATT

Series Title: Five Women Photographers; No. 5
25 min. col. video 1986
Great Britain English
Dir.: Judy Marle **Prod.:** Ruth Pantoleon
Prod. Agency: Broadside Productions; Channel Four Television
Foreign Distr: Albany Video (GB)

A profile of British photographer Helen Muspratt, featuring an interview with her. Five Women Photographers, No. 5 of 5.

HELMUT NEWTON

60 min. col. video; 16mm 1988
Federal Republic of Germany; Austria English (French and German language versions also avail.)
Dir.: Adrian Maben
Prod. Agency: RM Arts; ORF
Distributor (US): RM Associates/NY

Documents the work and the life-style of Australian photographer Helmut Newton (b.1920) in Paris, Berlin, Los Angeles, and Monte Carlo. Shows not only his art but also the business of his art and how he gathers about him the rich, the famous, and the beautiful, who are essential to his work. Newton discusses his childhood in Berlin and his innovations in the field of fashion photography. With contributions from Catherine Deneuve, Charlotte Rampling, Violetta Sanchez, Sigourney Weaver, Faye Dunaway, and Karl Lagerfeld.

Reviews: *Variety*, 15-21 Mar 89

HERITAGE FABULEUX DE NICEPHORE NIEPCE, L'

Series Title: Chroniques de France; No. 95
19 min. col. 16mm 1973
France French (English, Arabic, Portuguese, and Spanish language versions also avail.)
Dir.: Yves Allain
Prod. Agency: Telarci
Foreign Distr: French Ministry of Foreign Affairs (FR)
Foreign Distr: Association pour la Diffusion de la Pensée Française (FR) (Commercial & TV sales only)
Foreign Distr: Société Nouvelle Pathé Cinéma (FR)

French photographer Joseph Nicéphore Niépce (1765-1833) invented a method for making negative images using silver-sensitized materials. He experimented with fixing positive images by covering silver-coated copper plates with bitumen. After his death, Niépce's experiments were continued and a method for producing a direct positive image was discovered by his partner, French photographer Louis Jacques Mandé Daguerre (1787-1851). Chroniques de France series, No. 95.

HIROSHIMA BILDER

15 min. col. video 1983
Austria? German (French language version also avail.)
Dir.: Viktoria von Flemming
Foreign Distr: Goethe Institut, Lille (FR)

A series of photographs taken in Hiroshima, Japan, after the explosion of the first atomic bomb inspired Austrian painter, sculptor, and photographer Arnulf Rainer (b.1929).

HOCKNEY THE PHOTOGRAPHER

Series Title: South Bank Show, The
55 min. col. 16mm; video 1984
Great Britain English (French and German language versions also avail.)
Dir.: Don Featherstone; Jill Freeman
Prod. Agency: RM Arts; London Weekend Television
Distributor (US): RM Associates/NY
Distributor (US): Films Inc.
Distributor (US): artsAmerica
Distributor (US): Home Vision (Video sales only)
Distributor (US): Facets Multimedia (Video only)
Foreign Distr: London Weekend Television (GB) (Sales only)

Focuses on British painter David Hockney (b.1937), who has virtually stopped painting and has turned to the medium of photography. Hockney talks about this departure and is filmed at his Los Angeles home working on his *Joiner* photographs: photoassemblages showing different facial expressions and the passage of time. The South Bank Show series.

Reviews: *Los Angeles Daily News*, May 87; *Billboard*, 19 July 86; *Chicago Sun Times*, 1 Aug 86; *Video Insider*, 15 Sept 86; *The News-Herald*, 13 Feb 87; *Los Angeles Times*, 1 Apr 87

HOLLIS FRAMPTON

28 min. b&w video 1978
United States English
Prod. Agency: Video Data Bank
Distributor (US): Video Data Bank

Adele Friedman interviews American poet, photographer, and filmmaker Hollis Frampton (1936-1984), who discusses the development of his work, the influence of poet Ezra Pound, and the relationship between photography and filmmaking.

HOMAGE TO AUGUST SANDER
(HOMMAGE A AUGUST SANDER)

22 min. col. 16mm; video 1977
Federal Republic of Germany German with English subtitles
Dir.: Pavel Schnabel; Marlisa Stubenrauch
Prod. Agency: Pavel Schnabel Filmproduktion, FKR
Distributor (US): American Federation of Arts
Foreign Distr: German Film & Video Library (GB)

Profiles German photographer August Sander (1876-1964), who recorded the difficult lives of farm workers in a small farming town near Cologne during the early 1930s.

HORST

20 min. col. 16mm; video 1988
United States English
Dir.: Mark Trottenberg **Prod.:** Tina Barney
Exec. Prod.: Edgar B. Howard
Prod. Agency: Checkerboard Foundation, Inc.
Distributor (US): American Federation of Arts

Profiles German-born American photographer Horst P. Horst (b.1906), who traces his career as a fashion and portrait photographer in Europe and the United States. Recounts how he studied paintings at the Louvre in Paris for inspiration for his photographic compositions. Horst reminisces about working for *Vogue* magazine with publisher Condé Nast and editor Diana Vreeland. Includes interviews with former fashion model Lisa Fonssagrives-Penn and friend Valentine Lawford. Concludes with Horst conducting a fashion shoot for a lingerie advertisement.

IMAGE MAKERS

Series Title: Glory Trail, The
30 min. b&w 16mm 1964
United States English
Prod. Agency: NET
Distributor (US): Indiana University Audio Visual Center
Distributor (US): Rental-see *Educational Film & Video Locator*

Discusses the artists and photographers who portrayed the early American West and the two major sources that nurtured their efforts: the United States Government and the eastern newspapers. Shows works by artists John M. Stanley, Frederic Remington, Charles Marion Russell, and Cassilly Adams and photographers William H. Jackson and Timothy O'Sullivan. The Glory Trail series.

IMAGE OF LIGHT: THE PHOTOGRAPHY OF GEORGE POLLOCK

16 min. col. 16mm; video 1974
Great Britain English
Dir.: Anthony M. Roland **Prod.:** Anthony M. Roland

Prod. Agency: Anthony Roland Films
Distributor (US): The Roland Collection

Profiles photographer George Pollock (b.1928), who says: "For me, light is the most exciting thing there is....That's why my photographs are about light....The objects I put in front of my lenses are really only devices for controlling the light." He is vice-president of the Royal Photographic Society, serves on the council of the Royal Society of Arts, and is chairman of the London Salon of Photography and the "Photoeurop" Committee.

IMAGE OF THE CITY, THE

16 min. col. 16mm 1969p 1973r
United States English
Prod.: Charles Eames; Ray Eames
Distributor (US): Britannica Films
Distributor (US): Pennsylvania State University Audio Visual Services
Distributor (US): Rental-see *Educational Film & Video Locator*
Foreign Distr: Fergus Davidson Associates (GB)

Presents a series of historical photographs and images of cities. Using multiband cameras, radar, computer-generated graphics, satellite observations, and thermograms, examines ethnic housing problems, traffic, and crowds of the city. Based on an exhibition entitled "Photography and the City."

IMAGE, THE

Series Title: Exploring Photography; No. 6
25 min. col. video; 16mm 1978
Great Britain English
Prod.: Peter Riding
Prod. Agency: BBC Enterprises
Foreign Distr: BBC Enterprises (GB) (Sales only)

Discusses how photographers are exploring the limits of the medium in order to discover new ways of expressing themselves. Includes an interview with British photographer Paul Hill (b.1941). Exploring Photography series, No. 6 of 6.

IMAGES

Series Title: Images
6 programs, 30 min. ea. col. video 1989
Great Britain English
Prod. Agency: HTV Limited
Foreign Distr: HTV Limited (GB)

Presents the range of images that have been created during the 150 years since the invention of photography, using collections from Great Britain, North America, and France. Program 1 follows the technical and aesthetic advances made by the first photographers in the Victorian era, illustrated by examples from early photographic societies, including the Victoria and Albert Museum in London and the Royal Photographic Society of Great Britain in Bath. Program 2 demonstrates the scientific, archaeological, and anthropological uses of photography during the late Victorian period. Retraces the Middle Eastern travels of amateur archaeologist Gertrude Bell and visits sites in Jordan. Program 3 chronicles the emergence of photography as an art form, primarily in the United States, and features the work of American photographers Carleton E. Watkins, Alfred Stieglitz, and Edward Steichen. Program 4 examines the history of photography as a populist movement. Presents the work of street photographers Jimmy Forsythe from Tyneside and Jack Hulme from Castleford, whose works constitute documents of social history. Program 5 compares the worlds of the private collector and museum curator. Visits the collections of Michael Wilson in London, the Gilman Paper Company in New York, and the Library of Congress and National Museum of American Art in Washington, D.C. Program 6 examines photography as a commissioned art form and follows British photographer Christopher Killip (b.1946), winner of the 1989 Cartier-Bresson Award, to the Pirelli tire factory, where his assignment is to depict the factory at work. Images series.

IMAGES OF GLORY

Series Title: West of the Imagination, The
52 min. col. video 1986
United States English
Dir.: Tom Weidlinger **Prod.:** Tom Weidlinger
Exec. Prod.: David Kennard
Prod. Agency: KERA-TV in association with Independent Communications Associates (InCA)
Distributor (US): Films for the Humanities

Tells the story of the mythmakers of the American West and the myths they created: in oil or ink, still photographs and motion picture film. Surveys the historical and cultural landscape of the West. This episode chronicles the period of the Civil War, the last of the Indian Wars, and the driving of the final spike in the transcontinental railroad. Artists Thomas Moran (1837-1926) and Albert Bierstadt (1830-1902) paint a mythical Eden, while pioneer photographers Eadweard Muybridge (1830-1904) and Timothy O'Sullivan (1840-1882) whet America's appetite for western adventure. Narrated by actor James Whitmore. The West of the Imagination series, one of six.
Reviews: *EFLA Evaluations*, 87; *Sightlines*

IMOGEN CUNNINGHAM AT 93

13 min. col. 16mm; video 1978
United States English
Prod.: Irina Posner
Prod. Agency: CBS News
Distributor (US): Carousel Films
Distributor (US): Library of Congress (Archival only)

Portrait of American photographer Imogen Cunningham (1883-1976), who worked as a photographer when women were not accepted as professionals.

IMOGEN CUNNINGHAM, PHOTOGRAPHER

20 min. col. 16mm 1972
United States English
Dir.: John Korty
Prod. Agency: American Film Institute; Time-Life Films
Distributor (US): Pennsylvania State University Audio Visual Services
Distributor (US): Rental-see *Educational Film & Video Locator*

Documentary on the life and work of pioneer American photographer Imogen Cunningham (1883-1976), whose career, beginning in 1901 and continuing until her death, covered almost two-thirds the life span of photography itself. The artist, in her late eighties, is shown interacting with those around her and still spending several hours a day in her darkroom. Includes a retrospective look at her photographs from the early studies of nudes and flowers to later portraits and works using multiple exposures.
Awards: CINE Golden Eagle, 1974

IN CAMERA

30 min. col. 16mm 1984
Great Britain English
Dir.: Alistair Scott
Foreign Distr: Arts Council of Great Britain (GB)

Evokes the work of two Scottish pioneers of photography, David Octavius Hill (1802-1870) and Robert Adamson (1821-1848). Explores creativity and the complex technique of the equipment they used.

IN THE STREET

15 min. b&w 16mm 1952
United States Nonverbal
Dir.: Helen Levitt; James Agee; Janice Loeb
Distributor (US): Cecile Starr
Distributor (US): Pennsylvania State University Audio Visual Services
Distributor (US): Rental-see *Educational Film & Video Locator*

American photographer-filmmakers Helen Levitt, James Agee, and Janice Loeb document life among the poor in East Harlem, New York City, during the late 1940s. The filmmakers carried a camera hidden in a box, thus obtaining a candid portrait of these city dwellers.

INSTANT POUR L'ETERNITE, UN

Series Title: Histoire de la Photo
60 min. col. 16mm 1987
France French
Dir.: Florence Gruère **Prod.:** Claude Gallot
Prod. Agency: FR3
Foreign Distr: FR3 (FR)

The history of photography, focusing on Ansel Adams, Jacques-Henri Lartigue, Henri Cartier-Bresson, Robert Doisneau, André Kertész, and James Van Der Zee. Histoire de la Photo series.

INTERMITTENCES: E.J. MAREY 1830-1904

14 min. b&w 16mm 1978
France Nonverbal
Dir.: Jean-Marie Bouhours; M. Frizot
Prod. Agency: Centre Georges Pompidou
Foreign Distr: Centre du Film sur l'Art (BELG)

Shows photographs by French physiologist, inventor, and photographer Jules Etienne Marey (1830-1904). With his photographic research on movement, Marey stands with Eadweard Muybridge as one of the major figures in the history of late nineteenth-century photography. His inventions, the repeating-shutter camera and chronophotography, permitted him to fix the different positions of a body in motion through a series of consecutive images. Marey's work prefigures that of pioneer filmmakers, the Lumière brothers.

INVENTION DE LA PHOTOGRAPHIE, L'

21 min. col. 16mm 1965
France English (French, Spanish, Arabic, Portuguese, German, and Russian language versions also avail.)
Dir.: André Martin; Michel Boschet
Prod. Agency: Argos Films; Martin-Boschet
Foreign Distr: Argos Films (FR)
Foreign Distr: French Ministry of Foreign Affairs (FR)
Foreign Distr: Cinémathèque de l'Enseignement (FR)

Traces the steps of the discovery of photography in the nineteenth century and its subsequent expansion, paying tribute to the work of pioneer French photographer Joseph Nicéphore Niépce (1765-1833). Among other topics covered are heliography, French photographer Louis Jacques Mandé Daguerre (1787-1851) and the daguerreotype, the first negative prints on paper, the cinema, the beginnings of photojournalism, and microfilm.

INVENTION DE LA PHOTOGRAPHIE, L'

23 min. col. 16mm 1981
France French
Foreign Distr: Intermédia (FR)
Recounts the history of photography, highlighting the great discoveries of the nineteenth and twentieth centuries.

INVENTORS, THE

Series Title: History of World Photography, A
52 min. col. 16mm; video 1982
France English (French language version also avail.)
Dir.: Florence Gruère
Prod. Agency: RM Productions; FR3
Distributor (US): Producer Services Group (TV sales)
Foreign Distr: FR3 (FR)
Traces the invention of photography from its origins in the work of several different experimenters. Includes details of the contributions of Joseph Nicéphore Niépce, Louis Jacques Mandé Daguerre, Hippolyte Bayard, William Henry Fox Talbot, and others. Examines various processes they devised which advanced photographic technology. A History of World Photography series, one of four.

J.A. MARTIN, PHOTOGRAPHER

101 min. col. 16mm; video 1976
Canada English (French language version with English subtitles also avail.)
Dir.: Jean Beaudin **Prod.:** Jean-Marc Garand
Prod. Agency: National Film Board of Canada
Distributor (US): Karol Media
Foreign Distr: National Film Board of Canada (CAN)
Presents the fictional story of a woman of the end of the nineteenth century who leaves her children at home to accompany her husband on his yearly summer tour as an itinerant photographer. Follows their travels and depicts how they overcome fourteen years of mutual indifference to rediscover each other.
Awards: Cannes Palme d'Or, 1977; Cannes Best Actress, 1977; Cannes Special Jury prize, 1977; Film Advisory Board (Los Angeles) Award of Excellence, 1977; Canadian Film Awards Etrog, Best Screenplay, 1977; Canadian Film Awards Etrog, Best Director, 1977; Canadian Film Awards Etrog, Best Editing, 1977; Canadian Film Awards Etrog, Best Cinematography, 1977; Canadian Film Awards Etrog, Best Actress, 1977

J.L. SIEFF

Series Title: Chroniques de France; No. 143
7 min. col. 16mm; video 1977
France English (French, Arabic, Portuguese, and Spanish language versions avail.)
Foreign Distr: French Ministry of Foreign Affairs (FR)
Foreign Distr: Association pour la Diffusion de la Pensée Française (FR) (Commercial & TV Sales only)
Follows French photographer Jeanloup Sieff (b.1933) as he prepares an exhibition at the Centre Georges Pompidou in Paris. Chroniques de France series, No. 143.

JACQUES-HENRI LARTIGUE

58 min. col. 16mm; video 1985
France English
Dir.: Carl-Gustav Nykvist **Prod.:** Henrik von Sydow
Prod. Agency: Lucifer Films AB in cooperation with The Swedish Film Institute and Swedish Television
Distributor (US): FACSEA
Profiles French photographer Jacques-Henri Lartigue (1894-1986), who has been "writing" his diary with photographs since the turn of the century. Lartigue, commenting on this obsession, said, "I tried to capture everything." Includes photographs of his family, women's fashions of the 1920s, and auto races. Newsreel footage is used to place Lartigue and his work in their historical context.

JACQUES-HENRI LARTIGUE

Series Title: Master Photographers; No. 4
35 min. col. 16mm 1982
Great Britain English
Prod.: Peter Adam
Prod. Agency: BBC-TV; MM Chanowski Productions bv
Foreign Distr: Arts Council of Great Britain (GB)
Profiles French photographer Jacques-Henri Lartigue (1894-1986). His photograph albums, kept at the Grand Palais in Paris, constitute a diary of the twentieth century. He has been called the most deceptively simple and penetrating photographer in the history of the art. Master Photographers series, No. 4 of 6.

JACQUES-HENRI LARTIGUE, PEINTRE ET PHOTOGRAPHE

104 min. col. 16mm 1980
France French (English subtitled version also avail.)
Dir.: François Reichenbach
Prod. Agency: Les Films du Prisme; Antenne 2
Foreign Distr: Les Films du Prisme (FR)
Foreign Distr: Cinéculture (FR)
Foreign Distr: Intermédia (FR)

Explores the work of French painter and photographer Jacques-Henri Lartigue (1894-1986) in four episodes. Discusses his youth, ideas about artistic creation, career as a fashion and portrait photographer, and travels.

JACQUES-HENRI LARTIGUE: 1980

Series Title: Aujourd'hui en France; No. 24
5 min. col. 16mm; video 1980
France French (English, Spanish, and Arabic language versions also avail.)
Prod. Agency: French Ministry of Foreign Affairs
Distributor (US): Embassy of France
Foreign Distr: French Ministry of Foreign Affairs (FR)
Foreign Distr: ICS (FR)

Each day since the beginning of the twentieth century, French photographer Jacques-Henri Lartigue (1894-1986) has written his diary with pictures. Aujourd'hui en France series, No. 24.

J'AIME A LA FOLIE...LA PHOTOGRAPHIE

Series Title: J'aime la Folie
60 min. col. 16mm 1986
France French
Dir.: Michel Honorin
Prod. Agency: Antenne 2; Productions de la Chartreuse
Foreign Distr: France Media International (FR)
Foreign Distr: Antenne 2 (FR)
Foreign Distr: Productions de la Chartreuse (FR)

Visits the Arles International Photography Festival in France and explores the work of contemporary photographers represented there, including French photographers Jean-François Jonvelle and Robert Doisneau (b.1912). J'aime la Folie series.

JE TE PRENDS, TU ME PRENDS...EN PHOTO

Series Title: Je Te Prends, Tu Me Prends...En Photo
25 segments, 13 min. ea. col. video 1981
France French
Dir.: Madeleine Caillard
Prod. Agency: INF TV
Foreign Distr: INF TV (FR)
Foreign Distr: Madeleine Caillard (FR)

In a clear and concise vocabulary, presents the basis for a better mastery of photography techniques without ever losing the pleasure of looking. A series of twenty-five segments divided into five themes of five parts each: "Sharp and Soft Focus," "Equipment," "Light," "The Laboratory," and "Shooting and Framing." Je Te Prends, Tu Me Prends...En Photo series.

JEAN DIEUZAIDE: 1982

Series Title: Aujourd'hui en France; No. 74
5 min. col. 16mm; video 1982
France French (English, Spanish, and Arabic language versions also avail.)
Prod. Agency: French Ministry of Foreign Affairs
Distributor (US): Embassy of France
Foreign Distr: French Ministry of Foreign Affairs (FR)
Foreign Distr: ICS (FR)

An encounter with French photographer Jean Dieuzaide (b.1921) at the Arles International Photography Festival in France. Aujourd'hui en France series, No. 74.

JEANLOUP SIEFF

Series Title: Great Photographers
30 min. col. video 198?
Switzerland English?
Foreign Distr: Polivideo S.A. (SWIT)

Profiles French photographer Jeanloup Sieff (b.1933), best known for his nude portraits. Great Photographers series, one of five.

JETEE, LA

29 min. b&w 16mm 1962
France English (French language version also avail.)
Dir.: Chris Marker
Prod. Agency: Argos Films; Services de la recherche R.T.F.
Distributor (US): Biograph Entertainment, Ltd.
Foreign Distr: Intermédia (FR)

A dramatization constructed entirely of still photographs. Science-fiction story about a group of people who have survived an atomic bomb. They conduct time-travel experiments as a way of perpetuating the human race. One man journeys into the past and falls in love with a woman he had encountered as a child. He is killed by the other survivors.

JOHN ALBOK'S NEW YORK

Series Title: Eye on New York
30 min. b&w 16mm 1966
United States English
Dir.: Gordon Hyatt **Prod.:** Gordon Hyatt
Exec. Prod.: George Dessart
Prod. Agency: WCBS-TV
Distributor (US): Gordon Hyatt Inc. (Archival only)

Transforms five hundred photographs by Hungarian-born American photographer John Albok (1895-1982) into an animated interpretation of the spirit of New York City. A tailor by trade, Albok used his shop on Madison Avenue and East 96th Street as a darkroom at night to develop his pictures, many of them taken through his shop window. His photographs capture the human experience against the backdrop of New York City during the years 1925 to 1965. Eye on New York series.

JOHN BALDESSARI

Series Title: On Art and Artists
28 min. b&w video 1979
United States English
Prod.: Kate Horsfield; Lyn Blumenthal
Distributor (US): Video Data Bank

Nancy Bowen interviews American conceptual artist and photographer John Baldessari (b.1931), who discusses his work. On Art and Artists series.

JOHN BALDESSARI: A CONVERSATION

40 min. b&w video 1973
United States English
Distributor (US): Hermine Freed

American conceptual artist and photographer John Baldessari (b.1931) discusses his career and the development of his work. Presents examples of the artist's photographs and videos.

JOHN HEARTFIELD: PHOTOMONTAGIST (JOHN HEARTFIELD: FILM UEBER DEN FOTOMONTEUR)

63 min. col. 16mm; video 1976
Federal Republic of Germany English (German language version also avail.)
Dir.: Helmut Herbst **Prod.:** Helmut Herbst
Prod. Agency: Igelfilm GmbH
Distributor (US): American Federation of Arts
Foreign Distr: Igelfilm GmbH (BRD)

Profiles German painter, collagist, and filmmaker Helmut Herzfelde (1891-1968), who anglicized his name to John Heartfield in protest against the fanatical German chauvinism of World War I. About that time he turned from painting to photomontage, using it as a political weapon for antifascist statements during the 1920s and 1930s. A prominent member of the Berlin Dadaists, Heartfield contributed to publications such as *AIZ*, the German Communist Party weekly. Uses a variety of cinematic tricks to approximate Heartfield's photomontage technique and to give the viewer a sense of the artist's innovative polemical style.
Awards: German Film Critic's Award for Best Short Film, 1976

JOHN HOAGLAND: FRONTLINE PHOTOGRAPHER

29 min. col. video 1985
United States English
Prod.: David Helvarg
Distributor (US): Cinema Guild

Examines the life and work of American photojournalist John Hoagland (1952-1983), a specialist in war photography, who was killed in a firefight while working in El Salvador. Recalls his work for the Associated Press, United Press International, Gamma-Liaison, and *Newsweek* magazine. Includes interviews with Hoagland, friends, and family and excerpts from the feature film UNDER FIRE, whose central character was modeled after him. Shows many of Hoagland's photographs.
Reviews: *Choice*; *Library Journal*; *Artweek*

JOURNEE AVEC WILLY RONIS, UNE

11 min. col. 35mm; video 1984
France French
Dir.: Yves de Peretti
Prod. Agency: Centre Culturel Communal de Brétigny; Association Française pour la Diffusion du Patrimoine Photographique
Foreign Distr: Centre Culturel Communal de Brétigny (FR)

Follows French photographer Willy Ronis (b.1910) for a whole day in the Parisian neighborhood of Belleville.

JUST EVERGON

9 min. col. video 1988
Canada English with French subtitles
Dir.: Katherine Tweedie **Prod.:** Katherine Tweedie
Prod. Agency: National Gallery of Canada
Foreign Distr: Katherine Tweedie (CAN)

Foreign Distr: Cinéma Libre Inc. (CAN)
Presents Canadian photographer Evergon (b.1946), who produces work under the names of four different personas. Celluloso Evergonni uses large-scale Polaroid cameras to reenact Renaissance and baroque paintings; Eve R. Gonzalez, an eighty-year-old woman, makes memorial photographs in cemeteries; Egon Brut makes Polaroid studies in black and white from erotic photographs of the 1950s; and Evergon, who used to produce work in Xerox and cyanotype and now manages the other three personas. Describes the energy that characterizes his photographs and his efforts to work rapidly.

KARSH: THE SEARCHING EYE

79 min. col. 16mm; video 1985
Canada English
Dir.: Harry Rasky
Prod. Agency: Canadian Broadcasting Corp.
Distributor (US): Aylmer Press
Distributor (US): Baker & Taylor Video (Video sales only)
Foreign Distr: CBC Enterprises (CAN) (Sales only)
Profiles Canadian photographer Yousuf Karsh (b.1908), who emigrated from Armenia. Karsh discusses his photographic techniques in the making of a portrait of Leonard Bernstein and offers anecdotes about other notable subjects, including Winston Churchill, Eleanor Roosevelt, André Malraux, and Sophia Loren. Estrellita Karsh talks about meeting her husband and their life together.
Awards: International Film & TV Festival New York Silver Medal, 1986

KEEPERS OF THE LIGHT

Series Title: Great Experiments; No. 2
25 min. col. video 1988
Great Britain English
Dir.: George Auckland **Prod.:** Ron Bloomfield
Prod. Agency: BBC-TV
Foreign Distr: BBC Enterprises (GB) (Sales only)
Describes how key scientific experiments led to important technological developments in photography. Recreates in actual settings the experiments with light carried out by British photographer William Henry Fox Talbot (1800-1877) and French photographers Joseph Nicéphore Niépce (1765-1833) and Louis Jacques Mandé Daguerre (1787-1851). Great Experiments series, No. 2.

KEITH SMITH

19 min. b&w video 1976
United States English
Prod. Agency: School of the Art Institute of Chicago
Distributor (US): Video Data Bank
Interview with American photographer Keith Smith (b.1938), who discusses his combination of drawn and generated images in his pictures.

KENNETH JOSEPHSON

17 min. col. video 1983
United States English
Prod. Agency: Museum of Contemporary Art, Chicago
Distributor (US): Museum of Contemporary Art, Chicago
Chicago photographer and assemblage artist Kenneth Josephson (b.1932) discusses his early life and education, as well as his technique for making photographic series.

KENNETH SNELSON

Series Title: Video Vasari
10 min. col. video 1981
United States English
Prod. Agency: Albright-Knox Art Gallery
Distributor (US): Albright-Knox Art Gallery
A program in a series of video documents on art produced by the Albright-Knox Art Gallery, Buffalo, New York. Interviews American photographer and sculptor Kenneth Snelson (b.1927). Video Vasari series.

KENNETH SNELSON: THE PHOTOGRAPHS

12 min. col. video 1984
United States English
Prod.: Eric Shambroom; Robert Lewis
Distributor (US): DeCordova Museum and Sculpture Park
Produced in conjunction with an exhibition of panoramic photographs by American photographer Kenneth Snelson (b.1927), held at the DeCordova Museum and Sculpture Park, Lincoln, Massachusetts. Shows how Snelson uses an Eastman Cirkut camera to capture a 360-degree single image of New York City.

KLEINE BLEIBT KLEIN NICHT, DAS

30 min. col. video 1986?
Federal Republic of Germany German
Dir.: Alexander von Cube
Prod. Agency: WDR International
Foreign Distr: WDR International (BRD)
Documents formats of giant photography that permit a one-to-one size ratio between the reality filmed and the film itself. Explores the use of such techniques in the field of art restoration.

KODAK MUSEUM

12 min. col.& b&w video 1989
Great Britain English
Prod. Agency: BBC-TV
Foreign Distr: BBC Enterprises (GB)

Explores the Kodak Museum, devoted to the development of photography over the last 150 years, which opened in its new location in April 1989 at the National Museum of Photography, Film and Television in Bradford, England.

KODAK'S HOW TO TAKE BETTER PICTURES (BASIC PICTURE TAKING TECHNIQUES)

Series Title: Exploring Photography
30 min. col. video 1986
United States English
Prod. Agency: Eastman Kodak Co.
Distributor (US): Kodak Video Programs
Distributor (US): Baker & Taylor Video (Video sales only)
Distributor (US): Aylmer Press (Video sales only)

Introduces basic concepts of photography, including how to focus, centering the image, how and when to use a flash, and film speed. Explains lighting angles and how to capture a moving image by panning the camera. Uses photographs to illustrate the techniques. Exploring Photography series.

Reviews: *Landers Film Reviews*, Vol 32, No 2, Winter 87

L.E.A.R.: THE LEAGUE OF REVOLUTIONARY WRITERS AND ARTISTS

31 min. col. video 1988
Mexico? Spanish with English subtitles
Dir.: Yvette Nieves-Cruz **Prod.:** Yvette Nieves-Cruz
Distributor (US): Cinema Guild

Documents the history of The League of Revolutionary Writers and Artists (L.E.A.R.), founded in Mexico in 1934 by artists from various disciplines with shared political beliefs and a desire to support and serve workers through art and culture. Describes how during its four-year existence L.E.A.R. worked with union organizers, workers, and peasants to challenge the role of the United States in Latin America and the Caribbean. Surviving members of the league discuss their work and colleagues, including Diego Rivera, David Alfaro Siqueiros, Leopoldo Mendez, Pablo O'Higgins, and Tina Modotti. Incorporates archival footage and photographs, murals, paintings, and poems by L.E.A.R. members.

L.L. BEAN GUIDE TO OUTDOOR PHOTOGRAPHY

55 min. col. video 1985
United States English
Dir.: Bill Morris **Prod.:** Bill Morris
Distributor (US): Friendship II Productions
Distributor (US): Greenleaf Video Inc. (Video sales only)
Distributor (US): Baker & Taylor Video (Video sales only)

Outdoor photographer Lefty Kreh gives practical advice for improving 35mm color photography and covers the basics of light, composition, and equipment.

Reviews: *Choice*, Oct 86; *EFLA Evaluations*, 87
Awards: American Film & Video Festival Blue Ribbon, 1987

LANDSCAPE

Series Title: Exploring Photography; No. 3
25 min. col. video; 16mm 1978
Great Britain English
Prod.: Peter Riding
Prod. Agency: BBC Enterprises
Foreign Distr: BBC Enterprises (GB) (Sales only)

Deals with a wide variety of approaches to landscape photography, including the adventurous color work of Austrian photographer Ernst Haas (1921-1986) and the more symbolic black-and-white photographs of John Blakemore (b.1936). Exploring Photography series, No. 3 of 6.

LANDSCAPE PHOTOGRAPHER

11 min. col. video 198?
Great Britain English
Foreign Distr: BBC Enterprises (GB)

Discusses the work of British photographer Charlie Waite. Follows him on photographic projects in Scotland, France, and the Aran Islands off the coast of Ireland. Waite discusses how he selects his subjects and his love of landscape photography.

LANDSCAPE PHOTOGRAPHERS

Series Title: History of World Photography, A
52 min. col. 16mm; video 1982
France English (French language version also avail.)
Dir.: Florence Gruère
Prod. Agency: RM Productions; FR3
Distributor (US): Producer Services Group (TV sales)

Traces the contributions made to photography in the late nineteenth century by landscape photographers, who were often commissioned by the government to do surveys. A History of World Photography series, one of four.

LANGUAGE OF THE CAMERA EYE

Series Title: Photography — The Incisive Art
30 min. b&w 16mm; video 1962
United States English
Prod. Agency: KQED-TV; NET
Distributor (US): Indiana University Audio Visual Center
Distributor (US): Pennsylvania State University Audio Visual Services
Distributor (US): International Center of Photography (Archival only)

American photographers Ansel Adams (1902-1984) and Beaumont Newhall (b.1908) of George Eastman House analyze the photographs of Edward Weston (1886-1958), Henri Cartier-Bresson (b.1908), Edward Steichen (1879-1973), Alfred Stieglitz (1864-1946), and others. Adams explains the development of his own philosophy of photopoetry and how it has influenced his work. Photography — The Incisive Art series.

LAURA GILPIN: AN ENDURING GRACE

28 min. col.& b&w 16mm; video 1986
United States English
Dir.: Anita Thacher **Prod.:** William Howze
Prod. Agency: Amon Carter Museum
Distributor (US): American Federation of Arts
Distributor (US): Amon Carter Museum

Traces the life and work of American photographer Laura Gilpin (1891-1979), who documented the American Southwest and its native peoples for more than sixty years. Uses archival film footage, home movies (including excerpts from THE PLATINOTYPE DEMONSTRATED, a film Gilpin made with historian Beaumont Newhall to preserve a vanishing photographic process), interviews excerpted from Gretel Ehrlich's 1978 film LAURA GILPIN, as well as from other sources and a large selection of the artist's photographs to pay tribute to her long and productive life. Produced in conjunction with "An Enduring Grace: The Photographs of Laura Gilpin," an exhibition at the Amon Carter Museum in Fort Worth, Texas. Narrated by Gary Moody.

Reviews: *Sightlines*
Awards: CINE Golden Eagle, 1986; Sinking Creek Film Festival Award; San Francisco International Film Festival Honorable Mention

LAURIE SIMMONS: A TEASER

5 min. col. video 1980
United States English
Dir.: Carole Ann Klonarides; Michael Owen
Prod.: Michael Owen; Carole Ann Klonarides
Exec. Prod.: Jamie Davidovich
Prod. Agency: MICA-TV
Distributor (US): American Federation of Arts
Distributor (US): Video Data Bank

American photographer Laurie Simmons (b.1949) takes underwater photographs of swimming women, recalling the Weeki Wachee bathing beauties or the water ballets in Esther Williams films. The filmmakers taped Simmons through a glass window in the bottom of the swimming pool at Queens College in New York City while she was filming her models underwater. The underwater images of the swimmers transform the television set on which they are shown into a human aquarium.

LEISURE TIME PHOTOGRAPHY

20 min. col. 16mm; video 1980
Canada English
Dir.: David G. Harris
Prod. Agency: David G. Harris Productions Ltd.
Distributor (US): Beacon Films
Foreign Distr: David G. Harris Productions Ltd. (CAN)
Foreign Distr: Magic Lantern Films (CAN)

Canadian photographer Freeman Patterson (b.1937) conducts a course in photography, the art medium that he feels is the most accessible for leisure-time activities and that provides opportunities for self-expression.

Reviews: *Booklist*, 1 Nov 80; *Media & Methods Selected Short Films*, Dec 81
Awards: American Library Association Selected Films for Young Adults, 1982; Birmingham International Film Festival Electra Best in Category, 1981; Columbus International Film & Video Festival Chris Bronze Plaque, 1981; Chicagoland Educ. Film Festival Golden Statuette, 1981; Intl. Communication Film Competition, 1982

LENSES ON NATURE

27 min. col. 16mm 1981
Canada English
Dir.: Jack Carey **Prod.:** Jack Carey
Prod. Agency: Carey Productions Ltd.
Foreign Distr: Carey Productions Ltd. (CAN)

Shows the methods used by Maria and Len Zorn, two of the world's foremost nature photographers. Len specializes in bird photography, while Maria is especially interested in insects.

LEONARD MISONNE: THE PHOTOGRAPHER

58 min. col. 16mm; video 1979
Great Britain English
Dir.: Walter A. Franke; Ottoman Birth
Prod. Agency: RM Productions
Distributor (US): Producer Services Group (TV sales)

Takes a nostalgic journey into the past as recorded by the photographs of Belgian photographer Léonard Misonne (1870-1943). Because Misonne's career spans almost every important advance in photographic techniques, a survey of his work is at the same time a history of the medium.

"LET US NOW PRAISE FAMOUS MEN": ALABAMA 40 YEARS ON

35 min. col. 16mm 1979
Great Britain English
Dir.: Carol Bell **Prod.:** Alan Yentob
Prod. Agency: BBC-TV
Foreign Distr: Arts Council of Great Britain (GB)

In 1936 American writer James Agee (1909-1955) and American photographer Walker Evans (1903-1975) documented the deprivation and poverty of the Depression in Alabama. Their work was published in book form as *Let Us Now Praise Famous Men*, which has become a classic. Traces the survivors of the original families portrayed in intimate detail by Agee and Evans. Interweaves the families' memories and feelings with words and images from the book.

LICHFIELD ON PHOTOGRAPHY

Series Title: Lichfield on Photography
3 films, 50 min. ea. col. video 1982
Great Britain; Austria English
Dir.: Jim Goulding **Prod.:** Nick Scott
Exec. Prod.: Michael Norman
Prod. Agency: Original Image; EUMIG
Foreign Distr: Colstar Communications & Entertainment (GB)

A three-part series presented by British photographer Patrick Lichfield, designed to make learning about photography enjoyable as well as efficient. Covers basic tips on camera care, studio lighting, film formats, composition, lenses, darkroom activities, and photographing particular subjects such as children, landscapes, and sports. Lichfield on Photography series.

LIFE AND DEATH OF PICTURE POST, THE

60 min. col.& b&w 16mm 1977
Great Britain English
Prod.: John Ormand
Prod. Agency: BBC-TV
Foreign Distr: Arts Council of Great Britain (GB)

René Cutworth traces the history of *Picture Post*, a magazine that was a British institution during the years just before and after World War II and helped change the popular concept of the status of photography. The genius behind *Picture Post* was its creator and first editor Stefan Lorant, whose willingness to give free rein to young photographers led to the magazine's enormous success. Features the work of British photojournalist Bert Hardy (b.1913), whose pictures of American and South Korean atrocities during the Korean War led to a disagreement between the editor and owner and to *Picture Post*'s decline in the 1950s.

LITTLE BIT OF MAGIC REALISED, A

26 min. col. video 1986
Great Britain English
Dir.: Geoffrey Sax **Prod.:** Derek Clark
Prod. Agency: HTV West
Foreign Distr: HTV Limited (GB)

Sir Roy Strong, director of the Victoria and Albert Museum in London, examines the life of British photographer William Henry Fox Talbot (1800-1877), who invented the "photogenic drawing" (an early type of photograph).

LIVES OF LEE MILLER, THE

56 min. col. video 198?
Great Britain English
Foreign Distr: Dee and Co. Ltd. (GB)

Documents the life of American photographer, *Vogue* model, and war correspondent Lee Miller (1906-1947), who was also a lover and pupil of American painter and photographer Man Ray (1890-1976) and muse to artists such as Pablo Picasso, Georges Braque, Max Ernst, Paul Eluard, and Joán Miró. A legend in her own lifetime, she was a free and independent spirit with great talent.

LOCAL STRANGER, A

30 min. col. video 198?
Australia English
Dir.: Mark Shields-Brown **Prod.:** Mark Shields-Brown
Foreign Distr: SBS Television Marketing (AUSTRAL)

Profiles sixty-six-year-old Australian photographer Claude Beach, who lives alone out of a truck and travels to small towns taking portraits of residents. Shows Beach during his week-long stay in Julie Creek (population 450) located between Townsville and Mount Isa.

LOUIS JAMMES

Series Title: Lunettes Noires pour Nuits Blanches
? min. col. 16mm 1988
France French
Dir.: Dominique Colona **Prod.:** Catherine Barma; Thierry Ardisson
Prod. Agency: Antenne 2
Foreign Distr: Antenne 2 (FR)

Profiles French painter and photographer Louis Jammes (b.1958). Lunettes Noires pour Nuits Blanches series.

LUCIEN AIGNER: A LIFE WITH A CAMERA

29 min. col.& b&w video 1988
United States English
Dir.: Michael Ford **Prod.:** Nancy Norman
Exec. Prod.: Michael Ford; Mark Erstling (WGBY)
Prod. Agency: Yellow Cat Productions
Distributor (US): Yellow Cat Productions

Hungarian-born American photographer Lucien Aigner (b.1901) discusses his career. Focuses on his use of the Leica 35mm camera, introduced in the 1930s, which allowed the photographer to shoot in available light, thereby capturing the spontaneity of life. Shows his pictures of world political figures and celebrities, including Maurice Chevalier, Marlene Dietrich, Charles Laughton, Gypsy Rose Lee, and Charles Chaplin. Aigner recalls the events in his life that led him back to photography at the age of eighty-two.
Awards: Chicago Film & Television Festival Silver Hugo

LUMIERE PERE ET FILS

Series Title: Dossiers d'Alain Decaux, Les
? min. col. 16mm 1985
France French
Dir.: J.C. Dudrumet
Foreign Distr: Antenne 2 (FR)

Profiles French photographer Antoine Lumière (1842-1911) and his sons, pioneer filmmakers Auguste (1862-1954) and Louis (1864-1948). Les Dossiers d'Alain Decaux series.

MAN RAY

52 min. col.& b&w 16mm; video 1984
France English (French language version also avail.)
Dir.: Deidi von Schaewen; Heinz Schwerfel
Prod. Agency: Cineteve; Centre Georges Pompidou; Centre d'Action Culturelle d'Angoulême
Foreign Distr: Centre Georges Pompidou (FR)

Profiles American painter, photographer, and filmmaker Man Ray (1890-1976) with the use of archival photographs, film footage, and an extensive survey of his work in painting and photography. Excerpts from taped interviews of the artist provide the narration. Man Ray discusses the founding of the New York Dada group in 1917 by himself and French artists Marcel Duchamp and Francis Picabia. In 1921, Man Ray moved to Paris. He describes his activities in the Dada and surrealist movements there. The artist explains his photographic experimentation, which led to the creation of rayographs and solarizations. Shows an excerpt from his 1923 film, LE RETOUR A LA RAISON. Includes the on-screen participation of his wife, Juliet Man Ray, and of Teeny Duchamp, Meret Oppenheim, and Philippe Soupault.

MAN RAY

40 min. b&w video 1972
Great Britain English
Dir.: Peter Adam
Prod. Agency: BBC-TV
Foreign Distr: BBC Enterprises (GB)

Profiles American painter, photographer, and filmmaker Man Ray (1890-1976).

MAN RAY

Series Title: Plaisirs des Arts
10 min. b&w 16mm 1959
France French
Dir.: Michel Mitrani; Jacques Floran; Jean Luc Dejean
Prod. Agency: ORTF
Foreign Distr: INA (FR)

Profiles American painter, photographer, and filmmaker Man Ray (1890-1976). Plaisirs des Arts series.

MAN RAY PHOTOGRAPHE

Series Title: Chambre Noire
36 min. b&w 16mm 1961
France French
Dir.: Claude Fayard
Prod. Agency: ORTF

Foreign Distr: INA (FR)

Interviews American painter, photographer, and filmmaker Man Ray (1890-1976) in his studio in Paris. Rejected in the United States, where his paintings created a scandal, Man Ray emigrated to France. In Paris, his paintings were exhibited but didn't sell, and he turned to photography to earn a living. The artist discusses his photographs of celebrities and his use of the experimental techniques of solarization and rayographs. Chambre Noire series.

MANDEL/SULTAN

60 min. b&w video 1976
United States English
Prod. Agency: Video Data Bank
Distributor (US): Video Data Bank

Kate Horsfield interviews American photographers Mike Mandel (b.1950) and Larry Sultan (b.1946), who collaborate on billboard art projects under the name "The Clatworthy Colorvues."

MANUEL ALVAREZ BRAVO, FOTOGRAFO

29 min. col. 16mm 1983
Mexico Spanish
Dir.: Juan Mora
Prod. Agency: Universidad Nacional Autonoma de Mexico
Foreign Distr: Universidad Nacional Autonoma de Mexico (MEX)

Profiles Mexican photographer Manuel Alvarez Bravo (b.1904). Attempts to reconcile photography, a medium that arrests time and movement, and film, a medium characterized by movement.

MARCIA RESNICK

44 min. b&w video 1978
United States English
Prod. Agency: Video Data Bank
Distributor (US): Video Data Bank

Alex Sweetman interviews American photographer Marcia Resnick (b.1950), who discusses the range of her interests, from the nature of photography and its relation to art, to illusion, irony, and black humor. Talks about her recent fascination with punk culture.

MAREY: UN SAVANT TROP MECONNU

28 min. b&w 16mm 1976
France French
Dir.: P. Thévenard
Foreign Distr: Intermédia (FR)
Foreign Distr: French Ministry of Foreign Affairs (FR)

Profiles French physiologist, inventor, and photographer Jules Etienne Marey (1830-1904).

MARGARET BOURKE-WHITE

94 min. col. video 1989
United States English
Dir.: Lawrence Schiller **Prod.:** Lawrence Schiller
Exec. Prod.: Robert Halmi, Jr.; Rupert Dilnott-Cooper
Prod. Agency: Turner Network Television/TNT
Distributor (US): Turner Home Entertainment

Fictional biography of American photojournalist Margaret Bourke-White (1904-1971). Chronicles her career as a photographer for *Life* magazine beginning in 1936 and her marriage to author Erskine Caldwell, with whom she collaborated on several books of photographs. Depicts Bourke-White as she photographs Depression-era bread lines, General Patton at the front during World War II, and world leaders such as Mohandas K. Gandhi, Joseph Stalin, and Franklin D. Roosevelt. Stars Farrah Fawcett as Bourke-White and Frederic Forrest as Caldwell.

MARGARET MONCK

Series Title: Five Women Photographers; No. 3
25 min. col. video 1986
Great Britain English
Dir.: Judy Marle **Prod.:** Ruth Pantoleon
Prod. Agency: Broadside Productions; Channel Four Television
Foreign Distr: Albany Video (GB)

Profiles British photographer Margaret Monck, featuring an interview with her. She discusses her documentation of London's immigrant communities in the 1930s. Five Women Photographers, No. 3 of 5.

MARGINA 88

20 min. col. video 1988?
Yugoslavia Serbo-Croatian
Dir.: Sahin Sisic
Prod. Agency: JRT/Yugoslavia Radio-TV
Foreign Distr: JRT/Yugoslavia Radio-TV (YUGO)

While studying at the Zagreb Academy of Dramatic Arts in Yugoslavia, Czech photographer and filmmaker Sahin Sisic (b.1961) photographed the everyday life of the Lastavicas, a family with whom he occasionally stayed. Describes the moment when Sisic shows his photographs to the family and makes them leave their apartment and confront the outside world.

MARINE TERRACE

13 min. b&w 35mm; video 1988
France French
Dir.: Michel Pamart **Prod.:** Virginie Herbin
Prod. Agency: Musée d'Orsay; La Sept; Square Productions
Foreign Distr: Intermédia (FR)
Foreign Distr: Musée d'Orsay (FR)

Presents the daguerreotypes created by French author Victor Hugo (1805-1885) during his stay in exile on the isle of Jersey in the English Channel. Shows country scenes, family portraits, and self-portraits and discusses the projects that Hugo was contemplating: a new edition of his works illustrated with his photographs, massive distribution of his own portraits, and the publication of a work on the Channel Islands.

MARTIN CHAMBI AND THE HEIRS OF THE INCAS

Series Title: Arena
50 min. col.& b&w video; 16mm 1986
Great Britain Spanish with English subtitles
Dir.: Andy Harries; Paul Yule **Prod.:** Andy Harries; Paul Yule
Exec. Prod.: Anthony Wall; Nigel Finch
Prod. Agency: BBC-TV in association with Berwick Universal Pictures
Distributor (US): Cinema Guild
Distributor (US): Pennsylvania State University Audio Visual Services

Presents the life and work of Peruvian Indian photographer Martin Chambi (1891-1973), whose portraits documented both European and Native South American society. Features a multitude of Chambi's photographs, including those of the Inca ruins of Machu Picchu, and discusses their political, social, and artistic significance. Talks about Chambi's involvement in the "Indigenista" movement, which championed the indigenous Indian culture, and his photographic records of the remaining ancient Inca traditions. Discusses the plight of the Indians in contemporary Peru. Interviews include Chambi's daughters Mery and Celia; historians Atilio Sivirichi and José Luis Renique; Daniel Estrada, the current mayor of Cuzco, Peru; and several colleagues from the "Indigenista" group. Arena series.
Reviews: *Film & History*, Vol 17, No 3, Sept 87; *Library Journal*
Awards: American Film & Video Festival Blue Ribbon, 1988; Houston International Film Festival Gold Award; National Educational Film Festival Silver Apple; Society for Visual Anthropology Award of Excellence

MASS AGE IMAGE, THE: PHOTOJOURNALISM; PART I: THE BEGINNINGS OF PHOTOJOURNALISM: ERICH SALOMON; PART II: THE ESSENCE: TIM GIDAL

60 min. col. video 1982
United States English
Prod. Agency: Long Beach Museum of Art
Distributor (US): Long Beach Museum of Art
Distributor (US): International Center of Photography (Archival only)

Presents two segments on photojournalism. In Part 1, Els Barents discusses German photographer Erich Salomon (1886-1944), a pioneer in the field. Part 2 presents German photojournalist Tim N. Gidal (b.1909), who shows and discusses his work.

MASTER OF THE CAMERA: EDWARD J. STEICHEN

23 min. b&w 35mm; 16mm 1935
United States English
Prod. Agency: Pictorial Films
Distributor (US): Museum of Modern Art Dept. of Film (Archival only)

Shows American photographer Edward Steichen (1879-1973), a pioneer of modern photography, at work in his studio photographing a female model.

MATHEW BRADY: PHOTOGRAPHER OF AN ERA

12 min. b&w 16mm 1953p 1976r
United States English
Dir.: Lewis Jacobs **Prod.:** Paul Wendkos; Lewis Jacobs
Distributor (US): Public Media, Inc.
Distributor (US): Films Inc.
Distributor (US): Rental-see *Educational Film & Video Locator*

Biographical documentary of America's foremost pioneer photographer, Mathew B. Brady (1823-1896), emphasizing his work during the Civil War. In 1860, Brady was the most fashionable daguerreotype portrait photographer of his time. When the nation was suddenly divided by civil war, Brady left his galleries for the battlefront. He abandoned the daguerreotype for wet-plate photography, with its multiple prints; he trained the world's first combat crew, equipping his men with traveling darkrooms. He used the camera as it had never been used before: to record the history of a nation, to create a passionate social document. Includes photographs of American statesmen and presidents of the time and a portion of President

Abraham Lincoln's second inaugural address.
Awards: CINE Golden Eagle, 1976; Chicagoland Educ. Film Festival Golden Babe Award

MILTON ROGOVIN
Series Title: Video Vasari
12 min. col. video 1985
United States English
Prod. Agency: Albright-Knox Art Gallery
Distributor (US): Albright-Knox Art Gallery

A program in the series of video documents on art produced by the Albright-Knox Art Gallery, Buffalo, New York. Interviews American photographer Milton Rogovin (b.1909). Video Vasari series.

MILTON ROGOVIN
Series Title: Video Vasari
29 min. b&w video 1975
United States English
Prod. Agency: Albright-Knox Art Gallery
Distributor (US): Albright-Knox Art Gallery

A program in the series of video documents on art produced by the Albright-Knox Art Gallery, Buffalo, New York. American photographer Milton Rogovin (b.1909) talks about the aesthetic and ethical concerns of his socialist realist works. Video Vasari series.

MINUTE POUR UNE IMAGE, UNE
26 min. col. 35mm 1983
France French
Dir.: Agnès Varda
Prod. Agency: Garance; Centre National de la Photographie
Foreign Distr: Intermédia (FR)
Foreign Distr: Ciné-Tamaris (FR)

French filmmaker Agnès Varda presents a look at photographs and photography through eighteen one-minute episodes. Eighteen different people offer their personal interpretations of eighteen different photographs. Uses photographs by Deborah Turbeville, Luigi Ghirri, A. Martin, Gisèle Freund, P. Gérard Dupuy, Bernard Plossu, Josef Koudelka, Alain Resnais, Man Ray, Jacques Henri Lartigue, Agnès Varda, M. Barrat, Willy Ronis, Raymond Depardon, G. Polge, an anonymous photographer, and two photographs by Robert Doisneau.

MIROIR DE PAPIER, LE
12 min. col. 16mm 1969
France French (English, German, Spanish, Portuguese, Russian, and Polish language versions also avail.)
Dir.: Jean Vigne
Prod. Agency: Tybanis Film; Les Films Armorial
Foreign Distr: Tybanis Film (FR)
Foreign Distr: French Ministry of Foreign Affairs (FR)

Uses period photographs to trace the history of the medium, from its birth in the 1860s to the present day.

MODEL
129 min. b&w 16mm 1980
United States English
Dir.: Frederick Wiseman
Prod. Agency: Zipporah Films, Inc.
Distributor (US): Zipporah Films, Inc.

Shows male and female models working in television commercials and fashion shows and posing for magazine covers and print advertisements. Examines the work of photographers whose techniques illustrate different styles of fashion and product photography. Presents the business aspects of running a model agency, including interviewing prospective models, arranging portfolios, talking with clients, and planning trips.
Awards: CINE Golden Eagle, 1981

MOMENT IN TIME, A
Series Title: American Documents
53 min. col. 16mm; video 1976
United States English
Prod. Agency: PNS
Distributor (US): University of Illinois Film Center
Distributor (US): Rental-see *Educational Film & Video Locator*

American photographer Gordon Parks (b.1912) narrates this look at how the photographic image has become part of our lives and some of the people who made it happen. Photographers discussed include Mathew B. Brady, Ansel Adams, Dorothea Lange, Jacob Riis, Walker Evans, and Lewis W. Hine. The documentary value of the motion picture is also examined with clips from the films of Leni Riefenstahl, John Ford, Frank Capra, and John Huston. Shows excerpts from newsreels and mentions the contributions of Louis Jacques Mandé Daguerre, Edwin Herbert Land, and Thomas Alva Edison. Also mentioned is the impact of the 1888 Kodak camera. American Documents series, one of twelve.

MOMENT IN TIME, A

25 min. col. 16mm 1978
United States English
Distributor (US): National Audiovisual Center

Depicts the scope of the United States Navy's use of photography and its major photographic field activities.

MORLEY BAER

Series Title: Photoprofiles
30 min. col. video 1983
United States English
Dir.: John Morgan; Rebecca L. Morgan; Paul R. Schranz; Margaretta K. Mitchell **Prod.:** Thomas R. Schiff
Prod. Agency: Images Productions
Distributor (US): Images Productions

Profiles American photographer Morley Baer (b.1915), a specialist in architectural and landscape photography. Shows him at his home in Carmel, California, and at a shoot at Garrapata Beach on the Big Sur coast. He reminisces about growing up in Toledo, Ohio, his work as a commercial photographer in Chicago, and his desire to move to California. Baer also discusses his book in production, the culmination of his work photographing the California landscape. Photoprofiles series.

MOSCOU PARIS (1900-1930)

Series Title: Zig Zag
26 min. col. 16mm 1981?
France French
Dir.: J.M. Meurice **Prod.:** Teri Wehn Damisch
Prod. Agency: Antenne 2
Foreign Distr: Antenne 2 (FR)

Documents the opening of the "Moscow Paris" exhibition at the Pushkin Museum in Moscow. Discusses the Russian avant-garde. Zig Zag series.

MOVEMENT

Series Title: Exploring Photography; No. 1
25 min. col. video; 16mm 1978
Great Britain English
Prod.: Peter Riding
Prod. Agency: BBC Enterprises
Foreign Distr: BBC Enterprises (GB) (Sales only)

Discusses how photography can record movement as well as interpret it imaginatively. Includes an interview with Austrian photographer Ernst Haas (1921-1986), famous for his highly original color photographs. Exploring Photography series, No. 1 of 6.

MOVING STILL

Series Title: Nova
57 min. col. video 1981
United States; Great Britain English
Prod.: Christopher Hawes
Prod. Agency: WGBH; BBC-TV
Foreign Distr: BBC Enterprises (GB) (Sales only)

Traces the history of photography from the early camera obscura to the development of sophisticated techniques which are being used by scientists and engineers to capture extremely fast or slow movements. Presents Ernst Mach's first pictures of a bullet in flight and Dr. Harold Edgerton's invention of the stroboscope. Nova series.
Awards: American Film Festival Red Ribbon, 1982

MOVING STILLS

Series Title: Moving Stills
10 segments, 5 min. ea. col. video; 16mm 1989
Great Britain English
Dir.: Renny Bartlett **Prod.:** Renny Bartlett; Ian McAuley
Prod. Agency: Reverie Productions for the Arts Council of Great Britain in association with Channel Four Television
Foreign Distr: Arts Council of Great Britain (GB)

Presents the work of ten documentary still photographers living and working in Great Britain today. Each segment depicts the emotions and conflicts of its subject with accompanying music and minimal narration. THE OFFICE shows the humor, pathos, and irony of the workplace captured in photographs by Anna Fox. THE NEVER-ENDING POEM: CHILDREN OF THE BLACK TRIANGLE presents lyrical photographs of black children in the Americas, Great Britain, and Africa by Armet Francis. LOOKING OUT TO SEA: LIVERPOOL shows Peter Marlow's photographs of the people and their city. PILGRIMS documents cultural customs of mountain villages in Czechoslovakia recorded between 1966 and 1974 by Marketa Luskacova. NO EXPLANATION IS NECESSARY: THE NATIONALIST COMMUNITY OF NORTHERN IRELAND shows the daily life and pressures of that community in photographs by Mike Abrahams and Laurie Sparham. In BEIRUT: A KIND OF LIVING, Judah Passow charts a path from childhood to militiaman among the factions of a war-torn city. WAR EFFORT: WOMEN AT WAR examines the roles and responses of women to war in photographs by Jenny Matthews. In MARKET FORCES: TWO LONDON MARKETS — ONE MILE APART, Paul Trevor contrasts the financial markets of the City of London with the street market of Brick Lane in the East End. In ENGLISH PLEASURES, Chris Steele-Perkins captures the English at play. IN THE VOLCANO: CENTRAL AMERI-

CA uses dramatic action panoramas by Mike Goldwater to examine the nature of the struggle for power in Central America. Moving Stills series.

MUD OR BLOOD: ISSUES IN PHOTOJOURNALISM

25 min. col. video 1986
United States English
Dir.: John Schott
Prod. Agency: Walker Art Center
Distributor (US): Walker Art Center

Investigates the differences between art photography and photojournalism through interviews with photographers Jeff Jacobson, Alex Webb, and Jean-Marie Simon; picture editor Alice George; and photography historian Adam Weinberg, curator and author of *On The Line: The New Color Photojournalism.*

MUSIC IS THE STORY

4 min. b&w 16mm 1983
United States Nonverbal
Dir.: John Gati
Distributor (US): John Gati Film Effects, Inc.

Shows photographs by Hungarian-born American photographer André Kertész (1894-1985), accompanied by Béla Bartók's Suite op. 14 played by Frankel.

MUYBRIDGE IN MOTION

3 min. col. 16mm 1984
United States English
Dir.: Kenji Theil
Prod. Agency: University of Southern California, School of Cinema-TV
Distributor (US): University of Southern California Film & Video Distr. Ctr.

Multi-imaged study of photographs by British photographer Eadweard Muybridge (1830-1904), which increases in complexity from one man running to dozens of men, women, and animals in action.

Awards: CINE Eagle, 1984

MUYBRIDGE REVISITED

8 min. col.& b&w video 1986
Great Britain English
Dir.: George Snow
Prod. Agency: British Film Institute; Channel Four Television
Foreign Distr: British Film Institute (GB) (Archival only)

Uses computer-generated images to animate original photos by British photographer Eadweard Muybridge (1830-1904) and intercuts them with experimental video graphics and pictures of people and buildings in Fulham, an area of London. Muybridge's images of horses jumping, deer running, men wrestling, women dancing, and metal beaters are set in a roughly drawn street scene with ordinary but alienating architecture and accompanied by electronic pop music.

NADAR

? min. col. 16mm 1985
France French
Dir.: Gabriel Aloisi **Prod.:** Philippe Carrese; Gerard Detaille
Prod. Agency: FR3
Foreign Distr: FR3 (FR)

In the Canebière, Marseilles, an old shop keeps alive the memory of French photographer Nadar (Gaspard Félix Tournachon, 1820-1910).

NAKED EYE, THE

70 min. col. 16mm 1965
United States English
Dir.: Louis Clyde Stoumen
Distributor (US): Biograph Entertainment, Ltd.
Distributor (US): International Center of Photography (Archival only)
Distributor (US): Rental-see *Educational Film & Video Locator*

Traces the history of the camera and photography from early experiments by Leonardo da Vinci through the work of Mathew B. Brady, William Jackson, Louis Jacques Mandé Daguerre, Margaret Bourke-White, Alfred Eisenstaedt, Weegee (Arthur Fellig), and Edward Weston. Footage from Willard Van Dyke's 1948 film, THE PHOTOGRAPHER, shows Weston at work in California and Mexico.

Reviews: *New York Times*

NATHAN LYONS

44 min. b&w video 1978
United States English
Prod. Agency: Video Data Bank
Distributor (US): Video Data Bank

Alex Sweetman interviews American photographer and curator Nathan Lyons (b.1930), the founder and director of the Visual Studies Workshop in Rochester, New York. Lyons discusses his activities there and his involvement in validating photography's importance in education.

NEAL SLAVIN

60 min. b&w video 1977
United States English
Prod. Agency: Video Data Bank
Distributor (US): Video Data Bank

Joyce Neimanas interviews American photographer Neal Slavin (b.1941), who discusses his use of documentary photography in his books and prints.

NEVER GIVE UP — IMOGEN CUNNINGHAM

28 min. col. 16mm; video 1975
United States English
Dir.: Ann Hershey **Prod.:** Ann Hershey
Distributor (US): American Federation of Arts
Distributor (US): Phoenix Films & Video
Distributor (US): University of Illinois Film Center
Distributor (US): Buffalo State College Film Library
Distributor (US): International Center of Photography (Archival only)
Foreign Distr: Concord Films Council Ltd. (GB)

"There are some people who for some foolish reason want just me to photograph them. It's all foolishness," says American photographer Imogen Cunningham (1883-1976), whose pioneering career spanned over half the history of photography. Cunningham, whose work encompasses various styles of pictorialism, abstraction, and realism, has long been acclaimed for her portraits and her close-ups of plant forms. Shows her to be a charming and independent personality and presents an overview of her contribution to the history of photography.
Awards: American Film Festival Blue Ribbon, 1976

NEW YORK/NEW WAVE AT P.S. 1

Series Title: ART/new york; No. 5
28 min. col. video 1981
United States English
Dir.: Paul Tschinkel; Marc H. Miller
Prod. Agency: Inner-Tube Video
Distributor (US): Inner-Tube Video

A series on contemporary art that visits New York City galleries and museums and interviews artists and dealers. This segment shows work by American photographer Robert Mapplethorpe (1946-1989), current projects of graffiti artists, and excerpts from *Punk Magazine*. Includes interviews with artist Curt Hoppe and curator Diego Cortez. ART/new york series, No. 5.
Reviews: *Choice*, Jan 83; *Booklist*, Apr 81

NOSTALGIA

Series Title: Hapax Legomena; No. 1
36 min. b&w 16mm 1971
United States English
Dir.: Hollis Frampton
Distributor (US): Museum of Modern Art Circ. Film Library
Distributor (US): Film-Makers' Cooperative

An experimental film by American filmmaker Hollis Frampton (1936-1984) on the themes of time, memory, and anticipation. Shots of burning photographs are accompanied by a narrator's voice describing the content of the image yet to come. The photographs and narration concern incidents in Frampton's life. Hapax Legomena series, No. 1 of 7.
Awards: Bellevue Film Festival Grand Prize, 1971

OBJECTIF VILLE: ROBERT DOISNEAU

Series Title: Urba
? min. col. 16mm 1986
France French
Dir.: Jean Cazenave **Prod.:** Dominique Baillaud
Prod. Agency: FR3
Foreign Distr: FR3 (FR)

Observes the city through the eyes of French photographer Robert Doisneau (b.1912). Urba series.

OEIL NEUF, UN

Series Title: Société du 19ème Siècle et l'Aventure Photographique, La
14 min. col. 16mm; video 1973
France French (English, German, Russian, and Spanish language versions also avail.)
Dir.: Jean Vigne
Prod. Agency: Les Films Armorial; Tybanis Film
Foreign Distr: Intermédia (FR)
Foreign Distr: Tybanis Film (FR)
Foreign Distr: Les Films Armorial (FR)
Foreign Distr: French Ministry of Foreign Affairs (FR)

Chronicles the beginnings of photography in France from 1875 to 1900. La Société du 19ème Siècle et l'Aventure Photographique series.

OLIVER GAGLIANI

Series Title: Photoprofiles
30 min. col. video 1983
United States English

Dir.: John Morgan; Rebecca L. Morgan; Paul R. Schranz; Margaretta K. Mitchell **Prod.:** Thomas R. Schiff
Prod. Agency: Images Productions
Distributor (US): Images Productions

Profiles American photographer Oliver L. Gagliani (b.1917), who took up photography after a debilitating hearing loss impeded his music career. He discusses his introduction to photography; his early work; the influence of his teacher, Minor White; the metaphysical basis for imagery; and his recent color work in a documentary style. Photoprofiles series.

ONE SEEING...

15 min. col. 16mm 1971
United States English
Dir.: J.R. Holland
Prod. Agency: Christian Science Center
Distributor (US): Christian Science Center

Gordon Converse, chief photographer for *The Christian Science Monitor*, talks about his photographs for the newspaper and how he approaches his work.

OTHER OBSERVERS, THE: WOMEN PHOTOGRAPHERS FROM 1850 TO THE PRESENT DAY

72 min. col. video 1987
Great Britain English
Dir.: Brian Taylor
Prod. Agency: ILEA Learning Resources Branch
Foreign Distr: ILEA Learning Resources Branch (GB) (Sales only)

Documents women photographers, beginning with the amateur and professional photographers of the nineteenth century and concluding with work by contemporary women photographers. Photographers include Julia Margaret Cameron, Helen Muspratt, Dorothy Wilding, Grace Robertson, and Jill Posener.

OURSELVES AND THAT PROMISE

27 min. col. 16mm; video 1976
United States English
Dir.: Joe Gray, Jr.; Gene DuBey; Scott Faulkner
Prod. Agency: Appalshop Films
Distributor (US): Appalshop Films
Distributor (US): Rental-see *Educational Film & Video Locator*

Four Kentucky artists — writers Robert Penn Warren, James Still, and Ronnie Criswell and photographer Billy Davis — discuss the relationship between their work and their environment. The authors read from their works, and Warren discusses the work of American painter and ornithologist John James Audubon (1785-1851) while many of Audubon's prints are being shown. Displays Davis's photographs and accompanies the artist as he flies in a plane to capture aerial views of the Kentucky landscape.
Reviews: *Landers Film Reviews*

PARIS 09-31

28 min. col. video 1986
France French
Prod. Agency: Coll. Albert Kahn, Musée Départmental des Hauts-de-Seine
Foreign Distr: Coll. Albert Kahn, Musée Départmental des Hauts-de-Seine (FR)

Still photographs and films, juxtaposed chronologically, present a synthetic vision of the Albert Kahn collection of images of Paris and its streets.

PARIS 1985: LA GALERIE DE PHOTOS D'AGATHE GAILLARD

Series Title: Aujourd'hui en France; No. 150
4 min. col. 16mm; video 1985
France French (English, Spanish, and Arabic language versions also avail.)
Prod. Agency: French Ministry of Foreign Affairs
Distributor (US): Embassy of France
Foreign Distr: French Ministry of Foreign Affairs (FR)
Foreign Distr: ICS (FR)

Presents "The Month of the Photo," a huge festival spread throughout museums and galleries in Paris, with over one hundred exhibitions devoted to the best in photography. Highlights the photography gallery of Agathe Gaillard. Aujourd'hui en France series, No. 150.

PARIS NEW YORK

Series Title: Ping Pong
169 min. col. 16mm 1977
France French
Dir.: Pierre Desfons **Prod.:** Teri Wehn Damisch
Prod. Agency: Antenne 2; Centre Georges Pompidou
Foreign Distr: Antenne 2 (FR)

Describes the cultural and artistic exchange between Paris and New York since 1905. Ping Pong series.

PARIS OF ROBERT DOISNEAU, THE

20 min. col. 16mm 1973
France Nonverbal
Dir.: François Procile; Robert Doisneau

Prod.: Robert Doisneau
Prod. Agency: International Film Promotion
Distributor (US): FACSEA
Foreign Distr: International Film Production (FR)

A portrait of Paris over many years as viewed by French photographer Robert Doisneau (b.1912).

PATRICK ULLMANN, PHOTOGRAPHE DES CHANTEURS FRANCAIS

Series Title: Aujourd'hui en France; No. 90
5 min. col. 16mm; video 1982
France French (English, Spanish, and Arabic language versions also avail.)
Prod. Agency: French Ministry of Foreign Affairs
Distributor (US): Embassy of France
Foreign Distr: French Ministry of Foreign Affairs (FR)
Foreign Distr: ICS (FR)

Photographer of French popular singers and creator of numerous record covers, Patrick Ullmann concludes his series on performers with Georges Moustaki for an exhibition the Centre Georges Pompidou in Paris. Serge Reggiani, Raymond Devos, Barbara, and Claude Nougaro are among the featured stars in Ullmann's book *Top of the Bill*. Aujourd'hui en France series, No. 90.

PAUL LUTZ

Series Title: Chambre Noire
? min. col. video 1985
France French
Foreign Distr: FR3 (FR)

Profiles French photographer Paul Lutz, a specialist in aerial photography. Chambre Noire series.

PEINTURES FRAICHES, MADE IN FRANCE

52 min. col. 16mm 1985
France French
Dir.: Heinz Peter Schwerfel
Prod. Agency: Acmé Films
Foreign Distr: Acmé Films (FR)

Looks at the figurative art of the 1980s in France. Visits former warehouses in Vichy and sites in Paris where the camera follows and explores stairways and corridors, revealing wall paintings, broom sculptures, pictorial performances, photomontages, and works on old newspapers by Jean-Michel Albérola, Jean-Charles Blais, François Boisrond, Robert Combas, Louis Jammes, Denis Laget, Henri di Rosa, Georges Rousse, and Daniel Tremblay.

PETER LINDBERGH

Series Title: Great Photographers
30 min. col. video 198?
Switzerland English?
Foreign Distr: Polivideo S.A. (SWIT)

Observes German fashion photographer Peter Lindbergh on a location shoot in Normandy, France. Great Photographers series, one of five.

PHILIP TRAGER

Series Title: Photoprofiles
30 min. col. video 1983
United States English
Dir.: John Morgan; Rebecca L. Morgan; Paul R. Schranz; Margaretta K. Mitchell **Prod.:** Thomas R. Schiff
Prod. Agency: Images Productions
Distributor (US): Images Productions

Profiles American photographer Philip Trager (b.1935), who balances a career as a lawyer in Connecticut with his work as a photographer. He discusses how he maintains the creative energy necessary in both ventures. Shows Trager at his law office, at home, and on a shoot in Bridgeport, Connecticut. Photoprofiles series.

PHOTO ET LA MODE, LA

49 min. b&w 16mm 1967
France French
Dir.: J. Pradinas
Prod. Agency: ORTF
Foreign Distr: INA (FR)

Traces the evolution of fashion photography from its beginnings.

PHOTO JOURNALISME

50 min. b&w 16mm 1966
France French
Dir.: Jean Pradinas
Foreign Distr: INA (FR)

Press photographers are more than just illustrators of the events they witness — they are privileged observers of their eras. Includes French photographers R. Jammes, R. Martinez, Robert Doisneau, Roger Thirond, Daniel Camus, Jean Luis Swinners, and Henri Cartier-Bresson.

PHOTO SHOW, THE

Series Title: Photo Show, The
13 programs, 29 min. ea. col. video 1981

United States English
Prod. Agency: WGBH Educational Foundation
Distributor (US): PBS Video
A thirteen-part series demonstrating the basic skills and fine points of photography. Titles are ACCENTUATE THE NEGATIVE; A CHANGE OF LENS; FILTERS, DODGING AND BURNING; HUE AND EYE; IT'S ABOUT TIME; LET THERE BE LIGHT; MAKING CONTACTS; THE RIGHT EXPOSURE; STARTING TO PRINT; THE THREE KEY CONTROLS; TO CHOOSE A CAMERA; TOUGH SHOTS; and WHAT'S IN A FRAME? The Photo Show series.

PHOTO TIPS

10 parts, 3-8 min. ea. col. video 1986
United States English
Prod. Agency: Turrentine Productions
Distributor (US): Agency for Instructional Technology
A ten-part series in which commercial photographer Jerry Downs shares with viewers his ideas for creating outstanding pictures with simple equipment. He demonstrates specific solutions for common problems and reassures the novice that the best pictures come from taking time, preparing for the action, relaxing, and shooting. Examples of Downs's own work illustrate the many techniques he discusses. Program titles are BASEBALL, NIGHT, WEDDINGS, MACRO, VACATION, NATURAL LIGHT PORTRAITS, BABY, CARIBBEAN VACATIONS, ARCHITECTURE, and HUMOR.

PHOTO TRAVELLING

9 min. col. 16mm 1977
France French
Dir.: Alain Bellod
Prod. Agency: Centre Nat. de la recherche scientifique
Foreign Distr: Centre Nat. de la recherche scientifique (FR) (In France only)
Foreign Distr: Médiathèque Centrale (FR) (Non-France distr. only)
Explores the use of tracking to achieve photographic scale models of architectural structures.

PHOTOGRAPHE LASSINE, LE

6 min. col. 16mm 1973
France French
Dir.: Guy Marconnier
Prod. Agency: Films du Plateau
Foreign Distr: Inter Ciné TV (FR)
Presents an African photographer who uses a wooden box as a portable darkroom in which he makes identity card photographs.

PHOTOGRAPHE WILLY RONIS, LE

Series Title: Aujourd'hui en France; No. 68
5 min. col. 16mm; video 1981
France French (English, Spanish, and Arabic language versions also avail.)
Prod. Agency: French Ministry of Foreign Affairs
Distributor (US): Embassy of France
Foreign Distr: French Ministry of Foreign Affairs (FR)
Foreign Distr: ICS (FR)
In love with Paris, where he was born in 1910, French photographer Willy Ronis took numerous photographs of the city before crisscrossing the world as a top reporter. Now living in Provence, he is shown photographing the Isle-sur-Sorgue market. Aujourd'hui en France series, No. 68.

PHOTOGRAPHER OF THE YEAR

Series Title: National Geographic Explorer
18 min. col. video 1987
United States English
Prod.: Steve Burns
Exec. Prod.: Tim Kelly; Tom Simon
Prod. Agency: National Geographic Society
Distributor (US): Turner Broadcasting Services/WTBS
Award-winning National Geographic photographer Jim Stanfield (b.1938?) travels on assignment to Turkey to trace the footsteps of Ottoman emperor Süleyman the Magnificent (1494/95-1566). While photographing people in the Turkish countryside, Stanfield discusses his methods for capturing compelling moments on film. Shows him in Istanbul photographing the religious sect of whirling dervishes. Concludes with a display of Stanfield's choice photographs. National Geographic Explorer series.
Awards: CINE Golden Eagle, 1988

PHOTOGRAPHER, THE

Series Title: Arts-A-Bound!; Program 14
15 min. col. video 1985
United States English
Dir.: Gary Valente **Prod.:** Gary Valente; David A.Edmonds; Christine M. Murphy
Exec. Prod.: Thomas A. Valenti
Prod. Agency: WVIZ-TV; Educational Television Association of Metropolitan Cleveland

Distributor (US): Great Plains National Instructional Television Library

Presents a documentary field trip for sixth-, seventh-, and eighth-grade students to photographers' studios, with on-location interviews. Deals with the artistic decisions a photographer must make when taking a picture and developing the photograph. Discusses the similar elements and principles of visual art that are common to both photographers and painters. Photographer Brenda Lewison talks about her aesthetic choices, and Garie Crawford demonstrates new photographic technology as she uses color photocopying to create images. Arts-A-Bound! series, Program 14 of 16.

PHOTOGRAPHER, THE

Series Title: Accent
25 min. b&w 16mm 1961
United States English
Dir.: Martin Carr **Prod.:** Bill Kobin
Prod. Agency: CBS
Distributor (US): Library of Congress (Archival only)

American photographer Edward Steichen (1879-1973) discusses the development and concept of his work, defining himself as an incurable romantic. He explains his experimental approach to photography, describing some photographs in his 1961 retrospective exhibition at the Museum of Modern Art in New York City. Accent series.

PHOTOGRAPHER, THE

30 min. b&w 16mm; video 1948
United States English (French language version also avail.)
Dir.: Willard Van Dyke
Prod. Agency: Affiliated Film Producers for the United States Dept. of State
Distributor (US): Museum of Modern Art Circ. Film Library
Distributor (US): Biograph Entertainment, Ltd.
Distributor (US): National Audiovisual Center
Distributor (US): Pennsylvania State University Audio Visual Services
Distributor (US): Rental-see *Educational Film & Video Locator*
Foreign Distr: Ambassade des Etats-Unis (FR)

Portrait of pioneer American photographer Edward Weston (1886-1958). Considers Weston's work in interviews with him and with some of his friends. Weston talks about his work, the subject matter he chooses, and what it means to him. Discusses aspects of point of view, depth of field, and the process of selection involved in taking a picture. The film was directed by documentary filmmaker and photographer Willard Van Dyke.

PHOTOGRAPHER: RUSSELL LEE

58 min. col. video 1986
United States English
Dir.: Ann Mundy **Prod.:** Ann Mundy
Prod. Agency: Ann Mundy and Associates
Distributor (US): American Federation of Arts
Distributor (US): Ann Mundy and Associates

Profiles American social documentary photographer Russell Lee (1903-1986). A pioneer in flash photography, Lee recorded images of America during the Great Depression for the Farm Security Administration. Chronicles his series on coal miners for the Department of the Interior and his mapping of World War II air transport routes. Discusses the personal and artistic development of Lee as a photographer and his role in the history of social photography. Shows more than four hundred photographs and interviews American photographer Arthur Rothstein, former United States senator Ralph Yarborough, and curator Alan Fern.
Awards: CINE Golden Eagle, 1986; Texas Film Festival Best Documentary, 1986; National Educational Film & Video Festival Bronze Apple

PHOTOGRAPHER'S EYE, THE

Series Title: Creativity with Bill Moyers
29 min. col. video 1981
United States English
Prod. Agency: Corporation for Entertainment and Learning, Inc.
Distributor (US): PBS Video (Purchase only)

Explores the creative impulse in the art of photography. One photographer may see the same scene or object quite differently from another photographer. Examines the work of American photographers Emmet Gowin (b.1941) and Garry Winogrand (1928-1984) to show two very different approaches. Creativity with Bill Moyers series, one of seventeen.
Awards: Emmy Best Informational Series, 1982

PHOTOGRAPHERS OF EVERYDAY LIFE

Series Title: History of World Photography, A
52 min. col. 16mm; video 1982
France English (French language version also avail.)
Dir.: Florence Gruère
Prod. Agency: RM Productions; FR3
Distributor (US): Producer Services Group (TV sales)

Traces the evolution of contemporary techniques in photography used by French photographers Henri Cartier-Bresson (b.1908) and Brassaï (Gyula Halász, 1899-1984), and American photographer Ansel Adams (1902-1984) to document the world around them. A History of World Photography series, one of four.

PHOTOGRAPHERS OF THE AMERICAN WEST: 1860-1880

58 min. col. 16mm; video 1980
United States English
Dir.: Ray Witlin **Prod.:** Michael Blackwood
Prod. Agency: Blackwood Productions Inc.; Bayerischer Rundfunk; South Carolina Educational Television Network; Swedish Television-TV2
Distributor (US): Blackwood Productions Inc.

Features the work of the late-nineteenth-century photographers Carleton E. Watkins, Eadweard Muybridge, Andrew Russell, William Henry Jackson, Timothy O'Sullivan, William Soule, and John Hillers — the first group of photographers who traveled to the American West during the expansion and settling of the western territories from 1850 to 1890. Their photographs were widely disseminated in the eastern United States and in Europe, showing the public for the first time the natural wonders of the American West. Inspired by the exhibition "The Age of Exploration: The Rise of Landscape Photography in the American West, 1860-1885" held in 1975 at the Metropolitan Museum of Art in New York City.
Reviews: *Sightlines*, Spring 81

PHOTOGRAPHERS, THE

52 min. col. 16mm; video 1972
United States English
Prod. Agency: Time-Life Video
Distributor (US): Rental-see *Educational Film & Video Locator*

Observes five professional still photographers at work while on special assignments around the world: Larry Burrows, Alfred Eisenstaedt, John Dominis, Norman Parkinson, and Mary Ellen Mark.

PHOTOGRAPHES ET CRINOLINES

11 min. col. 35mm 1969
France French (English, Portuguese, Spanish, German, and Russian language versions also avail.)
Dir.: Jean Vigne
Prod. Agency: Les Films Armorial
Foreign Distr: Les Films Armorial (FR)
Foreign Distr: French Ministry of Foreign Affairs (FR)

Uses period photographs to trace the beginnings of photography and its role as witness to the Second Empire in France from 1852 to 1870.

PHOTOGRAPHIC DESIGN

Series Title: On Assignment
90 min. col. video 1987
United States English
Dir.: Gary Adams **Prod.:** Gary Adams
Exec. Prod.: Brian D. Ratty
Prod. Agency: Media West Home Video
Distributor (US): Media West Home Video (Video sales only)
Distributor (US): Baker & Taylor Video (Video sales only)

Host photographer Brian D. Ratty demonstrates tools, color arrangements, and compositional and lighting techniques that enhance photographic design. On Assignment series, one of seven.

PHOTOGRAPHIC LIGHT

Series Title: On Assignment
90 min. col. video 1986
United States English
Dir.: Ron Bourke **Prod.:** Gary Adams
Exec. Prod.: Brian D. Ratty
Prod. Agency: Media West Home Video
Distributor (US): Media West Home Video (Video sales only)
Distributor (US): Baker & Taylor Video (Video sales only)

Host photographer Brian D. Ratty discusses how the principles of utilizing light in photography offer more control and variety in the presentation of a subject. He gives a number of examples of how to arrange lights or use available light sources. On Assignment series, one of seven.

PHOTOGRAPHIC VISION, THE: A LIFETIME IN PHOTOGRAPHY

Series Title: Photographic Vision, The; Lesson 20
29 min. col. video 1984
United States English
Prod.: Sandra Ostertag
Exec. Prod.: Sandra Austin Harden
Prod. Agency: Coast Community Colleges in cooperation with KOCE-TV and CBS/Holt, Rinehart and Winston
Distributor (US): Coast Telecourses
Distributor (US): Library of Congress (Archival only)

A segment from a television course in photography. Features Max Yavno (1911-1985), an American photographer whose career has encompassed more than fifty years. His reflections on his career summarize many of the ideas presented throughout the course. The Photographic Vision series, Lesson 20 of 20.

PHOTOGRAPHIC VISION, THE: A VISUAL HERITAGE

Series Title: Photographic Vision, The; Lesson 2
29 min. col. video 1984
United States English
Prod.: Sandra Ostertag
Exec. Prod.: Sandra Austin Harden
Prod. Agency: Coast Community Colleges in cooperation with KOCE-TV and CBS/Holt, Rinehart and Winston
Distributor (US): Coast Telecourses
Distributor (US): Library of Congress (Archival only)

A segment from a television course in photography. Traces the highlights of the history of photography. American author and art historian Beaumont Newhall discusses the work of British portrait photographer Julia Margaret Cameron (1815-1879) and French photographer Nadar (Gaspard Félix Tournachon, 1820-1910). Edward W. Earle, curator of the Keystone Mast Collection at the California Museum of Photography, and George Eastman House curator Robert Sobieszek, describe other key figures and events. The Photographic Vision series, Lesson 2 of 20.

PHOTOGRAPHIC VISION, THE: ARTISTIC EXPRESSION

Series Title: Photographic Vision, The; Lesson 18
29 min. col. video 1984
United States English
Prod.: Sandra Ostertag
Exec. Prod.: Sandra Austin Harden
Prod. Agency: Coast Community Colleges in cooperation with KOCE-TV and CBS/Holt, Rinehart and Winston
Distributor (US): Coast Telecourses
Distributor (US): Library of Congress (Archival only)

A segment from a television course in photography. Looks at the artistic expression of two innovative American photographers, Nathan Lyons (b.1930) and Susan Rankaitis (b.1949). Robert Sobieszek, curator of the George Eastman House, discusses how the approach of Rankaitis is reflected in her highly individualistic images. The Photographic Vision series, Lesson 18 of 20.

PHOTOGRAPHIC VISION, THE: COLOR

Series Title: Photographic Vision, The; Lesson 10
29 min. col. video 1984
United States English
Prod.: Sandra Ostertag
Exec. Prod.: Sandra Austin Harden
Prod. Agency: Coast Community Colleges in cooperation with KOCE-TV and CBS/Holt, Rinehart and Winston
Distributor (US): Coast Telecourses
Distributor (US): Library of Congress (Archival only)

A segment from a television course in photography. Explores the reasons for the color revolution in fine-art photography which has occurred in the last ten years. American photographer Kenda North (b.1951) demonstrates her simplified version of the dye transfer color process and discusses the way color affects the aesthetics, dimensionality, and emotional content of a photograph. The Photographic Vision series, Lesson 10 of 20.

PHOTOGRAPHIC VISION, THE: CONTROLLING EXPOSURE

Series Title: Photographic Vision, The; Lesson 6
29 min. col. video 1984
United States English
Prod.: Sandra Ostertag
Exec. Prod.: Sandra Austin Harden
Prod. Agency: Coast Community Colleges in cooperation with KOCE-TV and CBS/Holt, Rinehart and Winston
Distributor (US): Coast Telecourses
Distributor (US): Library of Congress (Archival only)

A segment from a television course in photography. Photographers use examples of their work to discuss the art and craft of manipulating light in the camera. Examines the effect of shutter speed, aperture, and camera pans. American photographers John Divola (b.1949) and Linda Connor (b.1944) discuss their work and how they use the technical aspects of photography to achieve creative effects. The Photographic Vision series, Lesson 6 of 20.

PHOTOGRAPHIC VISION, THE: IMAGE AND IMPACT

Series Title: Photographic Vision, The; Lesson 1
29 min. col. video 1984
United States English
Prod.: Sandra Ostertag
Exec. Prod.: Sandra Austin Harden
Prod. Agency: Coast Community Colleges in cooperation with KOCE-TV and CBS/Holt, Rinehart and Winston
Distributor (US): Coast Telecourses
Distributor (US): Library of Congress (Archival only)

A segment from a television course in photography. Introduces the philosophy of the course: that photographic vision is a means of personal growth and expression. Previews program guests and concepts emphasizing three primary areas of professional photography: technical, commercial, and artistic. The Photographic Vision series, Lesson 1 of 20.

PHOTOGRAPHIC VISION, THE: LANDSCAPE AND CITYSCAPE

Series Title: Photographic Vision, The; Lesson 15
29 min. col. video 1984
United States English
Prod.: Sandra Ostertag
Exec. Prod.: Sandra Austin Harden
Prod. Agency: Coast Community Colleges in cooperation with KOCE-TV and CBS/Holt, Rinehart and Winston
Distributor (US): Coast Telecourses
Distributor (US): Library of Congress (Archival only)

A segment from a television course in photography. Traces the impact of the romantic movement in art on the landscape photography of American photographer Ansel Adams (1902-1984). American photographers Jack Welpott (b.1923), Linda Connor (b.1944), and Catherine Wagner also describe their approaches to their work. The Photographic Vision series, Lesson 15 of 20.

PHOTOGRAPHIC VISION, THE: PHOTOJOURNALISM

Series Title: Photographic Vision, The; Lesson 17
29 min. col. video 1984
United States English
Dir.: Sandra Ostertag **Prod.:** Sandra Ostertag
Exec. Prod.: Sandra Austin Harden
Prod. Agency: Coast Community Colleges in cooperation with KOCE-TV and CBS/Holt, Rinehart and Winston
Distributor (US): Coast Telecourses
Distributor (US): Library of Congress (Archival only)

A segment from a television course in photography. Photojournalist Ross Baughman describes his techniques for obtaining picture stories. Explores the issues of truth, objectivity, and ethics in photojournalism. The Photographic Vision series, Lesson 17 of 20.
Awards: CINE Golden Eagle, 1985

PHOTOGRAPHIC VISION, THE: READING PHOTOGRAPHS

Series Title: Photographic Vision, The; Lesson 8
29 min. col. video 1984
United States English
Prod.: Sandra Ostertag
Exec. Prod.: Sandra Austin Harden
Prod. Agency: Coast Community Colleges in cooperation with KOCE-TV and CBS/Holt, Rinehart and Winston
Distributor (US): Coast Telecourses
Distributor (US): Library of Congress (Archival only)

A segment from a television course in photography. American photographers Howard Becker and Nathan Lyons (b.1930) illustrate how visual communication carries more emotional, political, and cultural impact than does verbal communication. American photographers Michael Mandel (b.1950) and Larry Sultan (b.1946) discuss their individual and collaborative work. Warns the viewer to be aware of the effect of visual stimuli. The Photographic Vision series, Lesson 8 of 20.

PHOTOGRAPHIC VISION, THE: RESPONDING TO LIGHT

Series Title: Photographic Vision, The; Lesson 5
29 min. col. video 1984
United States English
Prod.: Sandra Ostertag
Exec. Prod.: Sandra Austin Harden
Prod. Agency: Coast Community Colleges in cooperation with KOCE-TV and CBS/Holt, Rinehart and Winston
Distributor (US): Coast Telecourses
Distributor (US): Library of Congress (Archival only)

A segment from a television course in photography. Helps students become more aware of the qualities and effects of available light. American photographer Jack Welpott (b.1923) takes the viewer to various indoor and outdoor settings to describe the way he responds to specific lighting conditions. Explores the emotional and aesthetic effects of light as well as the technical aspects. The Photographic Vision series, Lesson 5 of 20.

PHOTOGRAPHIC VISION, THE: SEEING WITH THE CAMERA

Series Title: Photographic Vision, The; Lesson 4
29 min. col. video 1984
United States English
Prod.: Sandra Ostertag
Exec. Prod.: Sandra Austin Harden
Prod. Agency: Coast Community Colleges in cooperation with KOCE-TV and CBS/Holt, Rinehart and Winston
Distributor (US): Coast Telecourses
Distributor (US): Library of Congress (Archival only)

A segment from a television course in photography. Demonstrates the technical differences between human and camera vision and discusses the personal vision of the photographer. Guest photographers Grant Mudford and Eileen Cowin describe and demonstrate their approaches to photographic interpretation. The Photographic Vision series, Lesson 4 of 20.

PHOTOGRAPHIC VISION, THE: STRUCTURE WITHIN THE IMAGE

Series Title: Photographic Vision, The; Lesson 9
29 min. col. video 1984
United States English
Prod.: Sandra Ostertag
Exec. Prod.: Sandra Austin Harden
Prod. Agency: Coast Community Colleges in cooperation with KOCE-TV and CBS/Holt, Rinehart and Winston
Distributor (US): Coast Telecourses
Distributor (US): Library of Congress (Archival only)

A segment from a television course in photography. America photographers Barbara Kasten (b.1936) and Garry Winogrand (1928-1984) show how every photographer uses a unique structure within his or her work. Discusses framing, line, shape, tone, and pattern and demonstrates ways to improve photographs structurally. The Photographic Vision series, Lesson 9 of 20.

PHOTOGRAPHIC VISION, THE: THE CAMERA

Series Title: Photographic Vision, The; Lesson 3
29 min. col. video 1984
United States English
Dir.: Sandra Ostertag **Prod.:** Sandra Ostertag
Exec. Prod.: Sandra Austin Harden
Prod. Agency: Coast Community Colleges in cooperation with KOCE-TV and CBS/Holt, Rinehart and Winston
Distributor (US): Coast Telecourses
Distributor (US): Library of Congress (Archival only)

A segment from a television course in photography. Photographic artist and educator Mark Johnstone uses a view camera to demonstrate and describe basic features which are common to all cameras. Illustrates the advantages and disadvantages of various types of cameras. American photographers Barbara Kasten (b.1936) and Garry Winogrand (1928-1984) explain how their particular cameras are ideal for their types of photography. The Photographic Vision series, Lesson 3 of 20.

PHOTOGRAPHIC VISION, THE: THE DARKROOM

Series Title: Photographic Vision, The; Lesson 11
29 min. col. video 1984
United States English
Prod.: Sandra Ostertag
Exec. Prod.: Sandra Austin Harden
Prod. Agency: Coast Community Colleges in cooperation with KOCE-TV and CBS/Holt, Rinehart and Winston
Distributor (US): Coast Telecourses
Distributor (US): Library of Congress (Archival only)

A segment from a television course on photography. American photographer Jerry Burchfield (b.1947) demonstrates the basic steps in developing and printing photographs. He discusses photograms, dodging, burning in, toning, solarization, and multiple printing. The Photographic Vision series, Lesson 11 of 20.

PHOTOGRAPHIC VISION, THE: THE FILM

Series Title: Photographic Vision, The; Lesson 7
29 min. col. video 1984
United States English
Prod.: Sandra Ostertag
Exec. Prod.: Sandra Austin Harden
Prod. Agency: Coast Community Colleges in cooperation with KOCE-TV and CBS/Holt, Rinehart and Winston
Distributor (US): Coast Telecourses
Distributor (US): Library of Congress (Archival only)

A segment from a television course in photography. Explores factors to be considered when selecting film for a specific photographic situation. Commercial photographer Jim Wood discusses the process of choosing the most appropriate film for an assignment. Explains how the speed of a film determines its graininess, contrast, suitability for action shots, and sensitivity to various lighting conditions. The Photographic Vision series, Lesson 7 of 20.

PHOTOGRAPHIC VISION, THE: THE MARKETPLACE

Series Title: Photographic Vision, The; Lesson 19
29 min. col. video 1984
United States English
Prod.: Sandra Ostertag
Exec. Prod.: Sandra Austin Harden
Prod. Agency: Coast Community Colleges in cooperation with KOCE-TV and CBS/Holt, Rinehart and Winston
Distributor (US): Coast Telecourses
Distributor (US): Library of Congress (Archival only)

A segment from a television course in photography. Dispels the notion that one cannot make a living in photography, but warns that creative approaches may be needed. David Fahey, assistant director of the G. Ray Hawkins photography gallery; Ellen Henderson, president of a photo agency; commercial photographer Jerry Abramowitz; and photojournalist Ross Baughman discuss their careers. Also looks at the career approach of American mountain climber and photographer Galen Rowell (b.1940). The Photographic Vision series, Lesson 19 of 20.

PHOTOGRAPHIC VISION, THE: THE PORTRAIT

Series Title: Photographic Vision, The; Lesson 14
29 min. col. video 1984
United States English
Prod.: Sandra Ostertag
Exec. Prod.: Sandra Austin Harden
Prod. Agency: Coast Community Colleges in cooperation with KOCE-TV and CBS/Holt, Rinehart and Winston
Distributor (US): Coast Telecourses
Distributor (US): Library of Congress (Archival only)

A segment from a television course in photography. Traces the tradition of portrait photography. Guests Arnold Newman (b.1918) and Judy Dater (b.1941), American portrait photographers with contrasting approaches, describe practical techniques for using lighting, pose, mood, and environment to create memorable photographic portraits. The Photographic Vision series, Lesson 14 of 20.

PHOTOGRAPHIC VISION, THE: THE STUDIO

Series Title: Photographic Vision, The; Lesson 12
29 min. col. video 1984
United States English
Prod.: Sandra Ostertag
Exec. Prod.: Sandra Austin Harden
Prod. Agency: Coast Community Colleges in cooperation with KOCE-TV and CBS/Holt, Rinehart and Winston
Distributor (US): Coast Telecourses
Distributor (US): Library of Congress (Archival only)

A segment from a television course in photography. Explains how a studio setting gives the photographer maximum control of all elements of the photograph. Jerry Abramowitz, a "product photographer," shows the photographer's responsibilities in a typical advertising assignment. American photographer Sandi Fellman (b.1952) discusses her work. The Photographic Vision series, Lesson 12 of 20.

PHOTOGRAPHIC VISION, THE: TIME AND MOTION

Series Title: Photographic Vision, The; Lesson 13
29 min. col. video 1984
United States English
Prod.: Sandra Ostertag
Exec. Prod.: Sandra Austin Harden
Prod. Agency: Coast Community Colleges in cooperation with KOCE-TV and CBS/Holt, Rinehart and Winston
Distributor (US): Coast Telecourses
Distributor (US): Library of Congress (Archival only)

A segment from a television course in photography. Examines the concept of "the decisive moment," defined by French photographer Henri Cartier-Bresson (b.1908) as the instant in which the technical and creative elements of photography come together to capture time and motion. Guest photographers discuss their approaches to action photography. The Photographic Vision series, Lesson 13 of 20.

PHOTOGRAPHIC VISION, THE: VISUAL DOCUMENTS

Series Title: Photographic Vision, The; Lesson 16
29 min. col. video 1984
United States English
Prod.: Sandra Ostertag
Exec. Prod.: Sandra Austin Harden
Prod. Agency: Coast Community Colleges in cooperation with KOCE-TV and CBS/Holt, Rinehart and Winston
Distributor (US): Coast Telecourses
Distributor (US): Library of Congress (Archival only)

A segment from a television course in photography. Edward W. Earle, curator of the Keystone Mast Collection at the California Museum of Photography, and American photographer Nathan Lyons (b.1930) trace the historical development of photography as a means of documentation and emphasize the importance of photographic records in fields such as science, industry, and law. Canadian-born American photographer Robbert Flick (b.1939) and American photographer David Scharf (b.1942) discuss their work. The Photographic Vision series, Lesson 16 of 20.

PHOTOGRAPHIE AMATEUR, LA

Series Title: Des Images et des Sons
14 min. col. 16mm 1981
France French
Dir.: D. Brillaud
Foreign Distr: Cinémathèque de l'Enseignement (FR)

Shows French photographer Lucien Clergue (b.1934) and his secondary school students at work in a photography lab. Des Images et des Sons series.

PHOTOGRAPHIE DEVANT LES BESOINS SPECIFIQUES DES ARCHEOLOGUES, LA

420 min. b&w video 1979
France French

Prod. Agency: Centre Nat. de la recherche scientifique
Foreign Distr: Centre Nat. de la recherche scientifique (FR) (In France only)
Foreign Distr: Médiathèque Centrale (FR) (Non-France distr. only)
Presents an international round table discussion on the different aspects of photography in archaeology.

PHOTOGRAPHIE ET SOCIETE

Series Title: Images de la Réalité
52 min. col. 16mm 1984
France French
Dir.: Teri Wehn Damisch **Prod.:** Gisèle Freund
Prod. Agency: Centre National des Arts Plastiques; TF 1
Foreign Distr: TF 1 (FR)
Foreign Distr: Cinéculture (FR)
Two-part program in which German-born French photographer Gisèle Freund (b.1912) presents her theories concerning photography and its use in understanding society. She illustrates her talk with many photographs and recounts the story of her flight from Nazi Germany. In Part 1, Freund discusses the photograph as a social weapon. In Part 2, she discusses photojournalism and the impact of news photographs on society. Images de la Réalité series.

PHOTOGRAPHIE, LA

6 min. col. 16mm 1984
Belgium? French
Dir.: Gerald Frydman
Prod. Agency: SODEP
Foreign Distr: French Ministry of Foreign Affairs (FR)
Recounts the invention of the cinematograph, showing that the conquest of the moving image was above all a masterpiece of formal ingenuity.

PHOTOGRAPHIE, LA: DONNER A VOIR

Series Title: Des Images et des Sons
14 min. col. 16mm 1981
France French
Dir.: D. Brillaud
Foreign Distr: Cinémathèque de l'Enseignement (FR)
Secondary school students work out a picture story and discover the problems posed by a pictorial narrative. The story is then photographed by a professional and an amateur photographer with different results. Des Images et des Sons series.

PHOTOGRAPHIE, LA: TECHNIQUES

Series Title: Des Images et des Sons
15 min. col. 16mm 1981
France French
Dir.: D. Brillaud
Foreign Distr: Cinémathèque de l'Enseignement (FR)
Shows French photographer Lucien Clergue (b.1934) and his secondary school students at the college of Tremblay studying the problems of photographic techniques. Examines format, automatic cameras, and zoom lenses and the different choices a photographer must make, including subjects, location, composition, lenses, and light. Des Images et des Sons series.

PHOTOGRAPHY

70 min. col. video 1989?
Canada English
Prod. Agency: Canadian Broadcasting Corp.
Foreign Distr: Canadian Broadcasting Corp. (CAN) (Archival only)
Explores the role that photography has played in society, emphasizing its documentary significance in the 150 years since its invention.

PHOTOGRAPHY

10 min. col. 16mm 1975
United States? English
Prod. Agency: Film D.R. Productions
Distributor (US): Reed Vocational Resource Library
Examines various careers in the field of photography, ranging from the medical photographer, the cinematographer, the news photographer, the portrait photographer, and the television cameraperson.

PHOTOGRAPHY

Series Title: Media Probes
30 min. col. 16mm; video 1982
United States English
Dir.: Mickey Lemle **Prod.:** Mickey Lemle
Prod. Agency: Laybourne/Lemle Inc.; Time-Life Video
Distributor (US): Buffalo State College Film Library
Focuses on five Americans who make their living behind the camera: photoessayist Bruce Davidson (b.1933), Pulitzer Prize winner David Kennedy, New York *Daily News* photojournalist Mary Dibiase, commercial photographer Michael O'Neill, and a photographer specializing in weddings, Gil Amaral. Intended to show the range of vision that is achieved through photography and the ways that it shapes our views of the world. Media Probes series.

PHOTOGRAPHY AS AN ART

Series Title: Photography — The Incisive Art
29 min. b&w video 1960
United States English
Prod. Agency: KQED-TV; NET
Distributor (US): Indiana University Audio Visual Center
Distributor (US): Pennsylvania State University Audio Visual Services
Distributor (US): International Center of Photography (Archival only)

Presents American photographer Ansel Adams (1902-1984) as he photographs Yosemite National Park in California. Shows a collection of his photographs and explains how he conveys a sense of discovery or rediscovery through his work. Adams discusses his methods of teaching photography and his debt to other photographers. Photography — The Incisive Art series.

PHOTOGRAPHY AS FINE ART

Series Title: Art Seen
28 min. col. video 198?
United States English
Dir.: David Howard
Prod. Agency: San Francisco Center for Visual Studies
Distributor (US): San Francisco Center for Visual Studies

Interviews American photographer Robert Heinecken (b.1931), who discusses his ideas on photography as an art form. Interviews museum curators and other photographers. Art Seen series.

PHOTOGRAPHY EXHIBITIONS

Series Title: ART/new york; No. 11
28 min. col. video 1982
United States English
Dir.: Paul Tschinkel; Marc H. Miller
Prod. Agency: Inner-Tube Video
Distributor (US): Inner-Tube Video

A series on contemporary art that visits New York City galleries and museums and interviews artists and dealers. This segment presents photography exhibits by American photographers Hans Namuth (b.1915) at the Leo Castelli Gallery, Cindy Sherman (b.1954) at Metro Pictures, and photographic works by American artist Robert Rauschenberg (b.1925) at the Sonnabend Gallery. Includes interviews with all three artists. ART/new york series, No. 11.
Reviews: *Choice*, Jan 83; *Booklist*, Apr 81

PHOTOJOURNALISM

15 min. col. video 1989?
Great Britain English
Foreign Distr: BBC Enterprises Library Sales (GB)

Profiles British photographer Brian Harris, who works for the *Independent* newspaper. Discusses how he manages to stay ahead of his competition.

PHOTOMONTAGE TODAY: PETER KENNARD

35 min. col. video 1982
Great Britain English
Dir.: Chris Rodrigues; Rod Stoneman
Prod.: Chris Rodrigues; Rod Stoneman
Prod. Agency: The Television Co-operative; Arts Council of Great Britain
Foreign Distr: Arts Council of Great Britain Film Library (GB)

Examines one tendency in the history of photomontage: to analyze the world in political terms. Looks at effects of layout and juxtaposition; the mechanics of photomontage construction and its efficacy; and the possible political uses for photomontage. Focuses on the work of British photomontagist Peter Kennard, who discusses the conception, construction, and distribution of his own work. Considers Kennard's work in relation to that of other photomontagists, such as John Heartfield and Hans Staek, who have used juxtaposition to make political statements.

PHOTOPROFILES

Series Title: Photoprofiles
31 min. col.& b&w video 1983
United States English
Dir.: Paul Schranz **Prod.:** Thomas R. Schiff
Distributor (US): Images Productions

Introductory program of the series Photoprofiles. Presents excerpts of interviews with the following photographers, as well as examples of their work: Ralph Steiner (landscapes); George Tice (industrial/architectural subjects); Ruth Bernhard (nudes); Eliot Porter (landscapes, nature); Oliver Gagliani (abstract); Betty Hahn (photomontage, manipulated colors); Philip Trager (architectural subjects); Willard Van Dyke (cinematography); Barbara Morgan (modern dancers); and Morley Baer (seascapes). Photoprofiles series.
Reviews: *Sightlines*, Summer 83

PHOTORAMA
14 min. col. 35mm 1978
Poland Polish
Dir.: Piotr Andrejew **Prod.:** Piotr Andrejew
Prod. Agency: Education Film
Foreign Distr: Film Polski (POL)
Profiles contemporary Polish photographer, painter, and graphic artist Zdzislaw Beksinski. Illustrates the story of his life with photographs of his country house.

PHOTOS D'ALIX, LES
15 min. col. 16mm 1980
France French (English subtitled version also avail.)
Dir.: Jean Eustache
Prod. Agency: Médiane
Foreign Distr: Médiane (FR)
As French photographer Alix lectures about her work, a young man becomes aware of the disparity between the artist's declared intentions and actual content of the pictures. The discussion, which is juxtaposed with the increasingly divergent images, is at odds with them and imposes meaning and interpretations not originally intended by the artist.

PHOTOSCOPIES
60 min. col. 16mm; video 1981
France French
Dir.: Erick Cloué
Foreign Distr: Cinéculture (FR)
Six French photographers meet and photograph six personalities from the art world: Michel Deluc and singer Christiane Eda Pierre; Jérôme Ducrot and accordionist Yvette Horner; François Le Diascorn and saxophonists Groupe Urban Sax; Guy Le Querrec and clarinetist Michel Portal; Jeanloup Sieff and dancer/choreographer Carolyn Carlson; and Martine Voyeux and sculptor/filmmaker Robert Lapoujade.

PICTURES
20 min. col. 16mm 1981
Canada English
Dir.: Robert McLachlan **Prod.:** Robert McLachlan
Exec. Prod.: Kurt Gagel
Prod. Agency: Omnifilm Productions Ltd.
Foreign Distr: Omnifilm Productions Ltd. (CAN)
Promotional film for a major photo lab that shows the entire consumer photo-finishing process without narration.

PICTURING DERRY
59 min. col. 16mm; video 1986
Great Britain English
Dir.: Dave Fox; Sylvia Stevens **Prod.:** David Glyn
Prod. Agency: Factions Films for the Arts Council of Great Britain in association with Channel Four Television
Distributor (US): Wombat Film & Video
Foreign Distr: Arts Council of Great Britain (GB)
Foreign Distr: Team Video Productions (GB) (Sales only)
Examines photographic images of Derry in Northern Ireland, a city associated with scenes of conflict and violence. International photojournalists Clive Limpkin, Ian Berry, and Terry Fincher discuss their experiences with taking pictures and getting them published.
Photographers from the Derry community Willie Carson, Willie Doherty, Barnie MacMonagle, Eileen Robson, and young members of Derry Camerawork are challenging what they feel is the distorted image of the city portrayed by outside photographers. A police photographer explains the official use of the camera. Argues that a photograph is more than just the picture it shows: the conditions under which it was taken, the context in which it is seen, and whether or not it is accompanied by a written text all influence the viewer.
Awards: American Film & Video Festival Red Ribbon, 1988

PIERRE MICHEL!!! ORCHESTRE A FICTIONS!!!
47 min. col. 16mm 1985
Belgium French
Dir.: Maddy Delsipée **Prod.:** Maddy Delsipée
Prod. Agency: Les Films du Chatelain; Centre Bruxellois de l'Audio-Visuel, RTBF
Foreign Distr: Centre du Film sur l'Art (BELG)
Profiles Belgian artist Pierre Michel (1919-1952) who committed suicide at age thirty-three, leaving behind a composite artistic achievement in painting, writing, photography, and film. Uses elements from all these areas to explore Michel's creative process.
Awards: Montreal Festival Intl. du Film sur l'Art Best Essay Award, 1985

PIONEERS OF PHOTOGRAPHY: COLOUR AND THE CAMERA
Series Title: Pioneers of Photography; No. 8
25 min. col. & b/w 16mm; video 1975
Great Britain English
Prod.: Ann Turner
Prod. Agency: BBC-TV

Foreign Distr: BBC Enterprises (GB)
Foreign Distr: Arts Council of Great Britain (GB)

Traces the development of color photography from the early 1860s to the first decade of the twentieth century and the gradual acceptance of photography as a popular art form. French chemist Louis Lumière (1864-1948) perfected the autochrome — a method of fixing three separate color layers to a glass slide for illumination by a projector — in 1907. The world soon went color mad said photographer Alfred Stieglitz. Shows the various color processes of the time, as well as the many ways in which color photography was used for book and magazine illustration. Pioneers of Photography series, No. 8 of 8.

PIONEERS OF PHOTOGRAPHY: FAMOUS MEN AND FAIR WOMEN

Series Title: Pioneers of Photography; No. 4
25 min. col. & b/w 16mm; video 1975
Great Britain English
Prod.: Ann Turner
Prod. Agency: BBC-TV
Foreign Distr: BBC Enterprises (GB)

Presents the life and work of British photographer Julia Margaret Cameron (1815-1879), who was introduced to the medium at the age of forty-eight. Her commentary and a demonstration of the cumbersome and time-consuming wet collodion process are combined to illustrate her dedication to the art of photography. Includes examples of her portraits and illustrations for Tennyson's *Idylls of the King*. Concludes with her journey to Ceylon, where she lived and made photographs until her death. Pioneers of Photography series, No. 4 of 8.

PIONEERS OF PHOTOGRAPHY: MIRROR WITH A MEMORY

Series Title: Pioneers of Photography; No. 2
25 min. col. & b/w 16mm; video 1975
Great Britain English
Prod.: Ann Turner
Prod. Agency: BBC-TV
Foreign Distr: BBC Enterprises (GB)

Documents the work of French photography pioneers Joseph Nicéphore Niépce (1765-1833) and Louis Jacques Mandé Daguerre (1787-1851), using spoken excerpts from their papers, as well as a wide selection of their photographs. Niépce created the first heliographs, images made directly through the action of light, and his experiments were continued after his death by Daguerre, who perfected the daguerreotype in 1839. Demonstrates the daguerreotype process. Pioneers of Photography series, No. 2 of 8.

PIONEERS OF PHOTOGRAPHY: NADAR THE GREAT

Series Title: Pioneers of Photography; No. 5
25 min. col. & b/w 16mm; video 1975
Great Britain English
Prod.: Ann Turner
Prod. Agency: BBC-TV
Foreign Distr: BBC Enterprises (GB)

Profiles French photographer Nadar (Gaspard Félix Tournachon, 1820-1910), who first used photography as an aid to his work as an illustrator and cartoonist. His mastery of light quickly became apparent in his portraits of famous friends — among them George Sand, Sarah Bernhardt, Charles Baudelaire, Jacques Offenbach, and Alexandre Dumas — as well as in his experiments with artificial light sources in the catacombs and sewers of Paris. Nadar survived his critics, who considered photography a lesser art form, and by the end of his life had become an elder statesman of the medium. Pioneers of Photography series, No. 5 of 8.

PIONEERS OF PHOTOGRAPHY: SUN PICTURES

Series Title: Pioneers of Photography; No. 3
25 min. col. & b/w 16mm; video 1975
Great Britain English
Prod.: Ann Turner
Prod. Agency: BBC-TV
Foreign Distr: BBC Enterprises (GB)

Records the achievements of three early photographers: Hippolyte Bayard (1801-1887) of France, and Robert Adamson (1821-1848) and David Octavius Hill (1802-1870), both from Scotland. Bayard used a technique of his own devising, as well as calotypes and daguerreotypes, to record the people and architecture of Paris in the 1840s. Hill, a painter, and Adamson, a chemist, formed a partnership in Edinburgh in 1848 and documented that city by means of their calotypes, or "sun pictures." Pioneers of Photography series, No. 3 of 8.

PIONEERS OF PHOTOGRAPHY: THE FLEETING IMAGE

Series Title: Pioneers of Photography; No. 7
25 min. col. & b/w 16mm; video 1975
Great Britain English
Prod.: Ann Turner
Prod. Agency: BBC-TV
Foreign Distr: BBC Enterprises (GB)

Examines the studies in motion carried out by British photographer Eadweard Muybridge (1830-1904) and French physiologist, inventor, and photographer Jules Etienne Marey (1830-1904). Muybridge pioneered the analysis of human and animal motion through sequential images, inventing a system of multiple cameras activated by trip wires. Marey invented several photographic methods for analyzing motion, including a "machine gun" camera for studying birds in flight, and the chronophotographic camera, a forerunner of the motion picture camera. Animation sequences demonstrate both men's contributions to scientific research of animal locomotion, as well as their importance to the beginnings of the cinema. Pioneers of Photography series, No. 7 of 8.

PIONEERS OF PHOTOGRAPHY: THE PENCIL OF NATURE

Series Title: Pioneers of Photography; No. 1
25 min. col. & b/w 16mm; video 1975
Great Britain English
Prod.: Ann Turner
Prod. Agency: BBC-TV
Foreign Distr: BBC Enterprises (GB)

Concentrates on the life and work of British pioneer photographer William Henry Fox Talbot (1800-1877). Shows the calotype process, by which Fox Talbot fixed a negative image on paper for the first time, for use in printing positive copies. Shows a selection of his work and includes spoken excerpts from his papers. Pioneers of Photography series, No. 1 of 8.

PIONEERS OF PHOTOGRAPHY: TRAVELLING MAN

Series Title: Pioneers of Photography; No. 6
25 min. col. & b/w 16mm; video 1975
Great Britain English
Prod.: Ann Turner
Prod. Agency: BBC-TV
Foreign Distr: BBC Enterprises (GB)

British photographer Samuel Bourne (1834-1912) arrived in India in 1863 and formed a partnership with Charles Shepherd. Their firm recorded the people, architecture, and landscape of the country. Demonstrates the wet collodion process Bourne used in his three expeditions to the Himalayas. Spoken excerpts from his writings accompany examples of his work: photographs of Moghul architecture, records of royal visits, and studies of racial types. Pioneers of Photography series, No. 6 of 8.

PIONNIERS, LES

Series Title: Histoire de la Photo
30 min. col. 16mm 1987
France French
Dir.: Bernard Bourgade **Prod.:** Michèle Arnaud
Prod. Agency: FR3
Foreign Distr: FR3 (FR)

Recounts the early history of photography, presenting pioneers in the field such as British photographer Eadweard Muybridge (1830-1904) and French photographers Maxime Du Camp (1822-1894) and Nadar (Gaspard Félix Tournachon, 1820-1910). Histoire de la Photo series.

PISTA: THE MANY FACES OF STEPHEN DEUTCH

28 min. col. 16mm; video 1979
United States English
Dir.: Katherine Deutch Tatlock **Prod.:** Katherine Deutch Tatlock; Andrew Hamersley
Distributor (US): Katherine Deutch Tatlock

Profile of Hungarian photographer and sculptor Stephen Deutch (b.1908). Through commentary, family photos, interviews, and extensive footage of his work, reveals his values, attitudes, and feelings about art, music, and his family. Features Deutch's wife, still photographer Helene Beck Deutch, who taught him about photography, as well as writers Studs Terkel and Nelson Algren and Fine Arts Quartet violinist Leonard Sorkin. Deutch has lived and worked in the Chicago area since 1936; he is known for his journalistic photographs featured in the *Daily News*, the *Sun Times*, and other publications.

Reviews: *Boston Globe*; *Eleven*, WTTW (Chicago PBS station) catalogue, Jan 89; *Chicago Magazine*, Nov 89
Awards: Great Lakes Film Festival Award, 1980

PLACE IN THE COUNTRY, A: LACOCK ABBEY, WILTSHIRE

Series Title: Place in the Country, A; No. 6
30 min. col. 16mm; video 19??
Great Britain English

Foreign Distr: National Trust Film Library (GB)
Visits Lacock Abbey in Wiltshire, England, a gothic fantasy house created in 1753 from a Tudor mansion which had originally been built using the remains of a thirteenth-century abbey. Lacock Abbey became the home and studio of British photographer William Henry Fox Talbot (1800-1877), who invented his original photographic processes there. Today, Lacock Abbey serves as the Fox Talbot Museum. A Place in the Country series, No. 6.

POINTS OF VIEW

Series Title: Photography — The Incisive Art
29 min. b&w 16mm; video 1960
United States English
Prod. Agency: KQED-TV; NET
Distributor (US): Indiana University Audio Visual Center
Distributor (US): Pennsylvania State University Audio Visual Services
Distributor (US): International Center of Photography (Archival only)
American photographer Ansel Adams (1902-1984) is shown photographing an old house and its inhabitants from many different perspectives and with many different purposes in mind. He illustrates how a tax assessor, a real estate agent, a decorator, a sociologist, a historian, a reporter, and a salon photographer might reflect the house in the lens. He then demonstrates how the house and its inhabitants can reveal themselves to the sensitive photographer and, through him, to a wide audience. Photography — The Incisive Art series.

POL MARA

13 min. col. 16mm; 35mm 1979
Belgium? French (Dutch language version also avail.)
Dir.: Marc Ghens
Foreign Distr: French Ministry of Foreign Affairs (FR)
The work of photographer and graphic artist Pol Mara, specialist in composition, reflects the sexuality and violence of his era.

PORTFOLIO

Series Title: Portfolio
10 programs, 4 min. ea. col. video 198?
Belgium? French?
Prod. Agency: Video Promotion; Clichés
Foreign Distr: Wallonie Bruxelles Images (BELG)
Clichés, a magazine dedicated to photography, presents ten programs, each focusing on a photographer and his world through his photographic art. Portfolio series.

PORTRAIT DE JEAN DIEUZAIDE

? min. col. 16mm 1986
France French
Dir.: Marc Christian
Prod. Agency: FR3
Foreign Distr: FR3 (FR)
Profiles French photographer Jean Dieuzaide (b.1921).

PORTRAIT D'UN PORTRAITISTE

5 min. col. video 1988
France French
Dir.: Laurent Sauerwein
Prod. Agency: R.I.P.
Foreign Distr: Films Grain de Sable (FR)
Reporter-photographer Jean-Marie Del Moral follows Spanish painter Miguel Barceló (b.1957). Visits the artist's studio in Paris and a church in Barcelona where he painted the cupola to show him in action. Del Moral discusses Barceló's work, which includes photographs taken in Africa where he resides today.

PORTRAIT OF IMOGEN

28 min. b&w 16mm; video 1987
United States English
Dir.: Meg Partridge **Prod.:** Meg Partridge; Nancy Hale
Distributor (US): New Day Films
Distributor (US): Pacific Pictures (Theatrical & TV rights)
Focuses on the life and work of American photographer Imogen Cunningham (1883-1976). Shows a range of photographs from her prodigious career from 1901 to 1976. The sound track is a compilation of recorded interviews of Cunningham taped primarily by her son, Rondal Partridge, and contains anecdotes about her personal and professional life.
Reviews: *Booklist*, 15 Mar 89; *Landers Film Reviews*, Summer 89; *Library Journal*, 15 May 89; *Choice*, May 89
Awards: Academy Award nomination for Best Documentary Short, 1988; Columbus International Film & Video Festival Chris Award, 1988; National Educational Film Festival Best of Northern California Award and Crystal Award, 1988; International Film & TV Festival of New York Bronze Award, 1988; CINE Golden Eagle, 1988; Focus Awards, First Place Documentary

PORTRAIT, THE

Series Title: Exploring Photography; No. 2
25 min. col. video; 16mm 1978
Great Britain English

Prod.: Peter Riding
Prod. Agency: BBC Enterprises
Foreign Distr: BBC Enterprises (GB) (Sales only)

The visual challenge of portrait photography is illustrated through the work of American photographer Arnold Newman (b.1918) and British photographers Duffy (b.1933) and Brian Griffin (b.1948). Exploring Photography series, No. 2 of 6.

PORTRAITS

Series Title: Painter's World, The: Changing Constants of Art from the Renaissance to the Present
28 min. col.& b&w 16mm; video 1988
United States English
Dir.: Judith Wechsler **Prod.:** Judith Wechsler
Exec. Prod.: Judith Wechsler
Prod. Agency: WGBH Educational Foundation; Channel Four Television; Judith Wechsler, Inc.
Distributor (US): Coronet/MTI Film & Video
Distributor (US): Producer Services Group (TV sales)

Explores the art of portraiture in both painting and photography, observing American painter Philip Pearlstein (b.1924) and American photographer Joel Meyerowitz (b.1938) at work. Describes conventions for representing persons of different social classes and professional groups. Explains how portraits became available to everyone with the invention of photography, yet lacked the subtle nuances of painted portraits. Narrated by the filmmaker, art historian Judith Wechsler. The Painter's World: Changing Constants of Art from the Renaissance to the Present series, one of six.

PORTRAITS

59 min. col.& b&w 16mm; video 1986
Switzerland French
Dir.: Luciano Rigolini **Prod.:** Luciano Rigolini
Prod. Agency: Radiotelevisione della Svizzera Italiana; SSR
Foreign Distr: Luciano Rigolini (SWIT)
Foreign Distr: Radiotelevisione della Svizzera Italiana (SWIT)

Profiles ten contemporary Swiss artists: painters John Armleder, Adriana Beretta, Martin Disler, Helmut Federle, and Peter Fischli; graphic artist Miriam Cahn; photographers David Weiss and Hannah Villiger; sculptor and lithographer Josef Felix Müller; and video and performance artist Ann Winteler. Each portrait was done in close collaboration with the artist and is in harmony with the works being presented.

Awards: Montreal Festival Intl. du Film sur l'Art Best Film for Television, 1987; Festival International du Film d'Art et d'Archéologie de Bruxelles Grand Prix, 1988

PORTRAITS AND DREAMS

18 min. col. video 1985
United States English
Prod. Agency: Appalshop Films
Distributor (US): Appalshop Films

Presents the photographs of children taught by Wendy Ewald, director of Appalshop's Mountain Photography Workshop. The children, from several mountain schools in Letcher County, Kentucky, were instructed to photograph the world around them. The resulting images became a touring exhibition produced by the Smithsonian Institution. Includes recorded interviews with the children and Ewald.

Reviews: *Booklist*

PORTRAITS ET GRIMACES: WITKIEWICZ

30 min. col. video 1989
France French
Dir.: Jean-Pierre Krief
Prod. Agency: La Sept; KS Visions
Foreign Distr: KS Visions (FR)
Foreign Distr: La Sept (FR)

Surveys the photographic work of Polish painter and author Stanislaw Ignacy Witkiewicz (1885-1939). From the age of eight until his death, he photographed his family, friends, and himself. His portraits are a psychological exploration of human nature: scared, wicked, and innocent faces. Uses excerpts from Witkiewicz's writings to probe the meaning of his work.

PREHISTORY OF CINEMA, THE

Series Title: Prehistory of Cinema, The
13 programs, 12 min. ea. col. video 198?
Belgium English?
Foreign Distr: SODEP (BELG)

A thirteen-part series that combines documentary techniques with recreations to recount the prehistory of cinematography. Discusses the contributions of Gaspard Robertson; Joseph Plateau, inventor of "the revolving window," the future cinematographic shutter; Emile Reynaud, inventor of the praxinoscope in 1877; Louis Jacques Mandé Daguerre (1787-1851), inventor of the daguerreotype; Eadweard Muybridge (1830-1904); Jules Etienne Marey (1830-1904); Louis Aimé Le Prince; William Friese-Green; and Thomas Alva Edison (1847-1931), who patented the kinetoscope. The Prehistory of Cinema series.

PRIMITIFS DE LA PHOTOGRAPHIE, LES

Series Title: Zig Zag
26 min. col. 16mm 1981
France French
Dir.: Pierre Desfons **Prod.:** Teri Wehn Damisch
Prod. Agency: Antenne 2
Foreign Distr: Antenne 2 (FR)
Presents the pioneers of photography. Zig Zag series.

PROCEDE FRESSON, LE

Series Title: Procédés, Les
30 min. col. 16mm 1987
France French
Dir.: Jean Réal **Prod.:** Pierre Le Bret
Prod. Agency: Artistes et Artisans; Cinq Continents; Centre Nat. de la recherche scientifique; Centre National des Arts Plastiques; Rencontres Internationales de la Photographie d'Arles; French Ministry of Foreign Affairs; Intermédia
Foreign Distr: Intermédia (FR)
Foreign Distr: Cinq Continents (FR)
Foreign Distr: Centre Nat. de la recherche scientifique (FR)

The Paris studio of Michel and Jean François Fresson is the only studio in the world that uses the carbon-printing process developed by the Fresson family at the end of the nineteenth century. Although the principles of the process have remained unchanged, the monochrome-based technique was adapted for color printing in 1952. The prints have proved very stable and resistant to degradation over time. The Fresson Process allows a wide latitude for manipulation and variation of results. It can produce prints that are very realistic as well as prints that achieve a dimension of imagination, such as the photographic experiments of the Pictorialists. Includes the participation of Bernard Faucon, John Batho, Bernard Plossu, Frank Horvat, and Georges Tourdjman. Les Procédés series, one of three.

Awards: Montreal Festival Intl. du Film sur l'Art, Special Jury Prize, 1988

PROFESSIONAL PHOTOGRAPHY

Series Title: Photography — The Incisive Art
29 min. b&w 16mm; video 1960p 1980r
United States English
Prod. Agency: KQED-TV; NET
Distributor (US): Indiana University Audio Visual Center
Distributor (US): International Center of Photography (Archival only)

American photographers Ansel Adams (1902-1984) and Milton Halberstadt discuss photography as a profession and explain the various techniques used in commercial, industrial, promotional, and portrait photography. Photography — The Incisive Art series.

QUAI BOURBON

Series Title: Découverte d'une Oeuvre
6 min. b&w 35mm; video 1986
France Nonverbal
Dir.: Luc Riolon
Exec. Prod.: Jean-Pascal Levy-Trumet; Denis Bancilhon
Prod. Agency: Musée d'Orsay
Foreign Distr: Musée d'Orsay (FR)

The arts of cinema, dance, and music combine in this visual essay inspired by the photographs *Les Ramoneurs* and *Quai Bourbon* (1852) by French painter and photographer Charles Négre (1820-1880). Découverte d'une Oeuvre series.

QUEBEC AS SEEN BY CARTIER-BRESSON, LE
(QUEBEC VU PAR CARTIER-BRESSON, LE)

10 min. b&w 16mm; video 1969
Canada English (French language version also avail.)
Prod.: Guy Glover
Prod. Agency: National Film Board of Canada
Foreign Distr: National Film Board of Canada (CAN)

French photographer Henri Cartier-Bresson (b.1908) turns his lens on the citizens of Quebec, capturing the form and movement that characterize his work. In town and country, he photographs young people, old people, streets and fences, homes and edifices in discrete moments which, when assembled, provide a microcosm of Quebec.

QUELLE BELLE EPOQUE

10 min. col. 35mm; 16mm 1981
France French
Dir.: Stanislas Choko
Prod. Agency: Iskra
Foreign Distr: Iskra (FR)

Uses picture postcards from the Belle Epoque (1890-1914) in France to juxtapose its idyllic aspects (represented by actress Sarah Bernhardt, early aviation, and café nightlife) with the tougher realities of the time.

RALPH STEINER

Series Title: Photoprofiles
30 min. col. video 1983
United States English

Dir.: John Morgan; Rebecca L. Morgan; Paul R. Schranz; Margaretta K. Mitchell **Prod.:** Thomas R. Schiff
Prod. Agency: Images Productions
Distributor (US): Images Productions
Profiles American photographer and filmmaker Ralph Steiner (1899-1986). He talks about his beginnings in photography, his ideas on photographic education, and his relationship with Willard Van Dyke, Paul Strand, and other documentary photographers. Photoprofiles series.

RAPHO, HISTOIRE D'UNE FAMILLE

40 min. col.& b&w 35mm 1987
France French
Dir.: Frédéric Mitterand; Patrick Jeudy
Prod. Agency: Rapho
Foreign Distr: Intermédia (FR)
A film commissioned by the Arles International Photography Festival for the fiftieth anniversary of the Rapho agency. Constructed in the form of a private diary, the author evokes a slice of life and history combining emotions and autobiographical souvenirs with the memories associated with the photographs.

REAL WEEGEE, THE

58 min. col.& b&w video 1965p 1987p
United States English
Dir.: Sherman Price
Prod. Agency: Sherman Price Productions
Distributor (US): Baker & Taylor Video (Video sales only)
Combines two films about American photographer Weegee (Arthur Fellig, 1900-1968). The first, a fabricated autobiography made in 1965, follows Weegee as he pursues his unrequited love, a mannequin, across Europe. The second, a documentary profile made in 1987, examines more than one hundred photographs from Weegee's bestselling books *The Naked City* and *Weegee's People.*
Reviews: *Video Librarian,* Vol 4, No 7, Oct 89

REALISME ET ILLUSION

Series Title: Histoire de la Photo
60 min. col. 16mm 1987
France French
Dir.: Patrick Jeudy **Prod.:** Michèle Arnaud
Prod. Agency: FR3
Foreign Distr: FR3 (FR)
Looks at the essential themes of twentieth-century photography: fashion, the portrait, and critical expression. Histoire de la Photo series.

RECITS D'ELLIS ISLAND: TRACES

58 min. col.& b&w 16mm 1980
France French
Dir.: Robert Bober; Georges Perec
Prod. Agency: INA
Foreign Distr: Intermédia (FR)
A recollection of Ellis Island through the pictures of American photographer Lewis W. Hine (1874-1940). From 1892 to 1924, almost sixteen million European immigrants, driven out of their homelands by poverty, hunger, and political, religious, or racial oppression, fled to America. Evokes the memory of the emigrés passing through the reception center at Ellis Island, an islet near the Statue of Liberty in New York harbor.

REFLECTIONS

52 min. col. video 1988
Federal Republic of Germany English (German language version also avail.)
Dir.: Gordon Swire **Prod.:** Friedmann Beyer; Cynthia King
Exec. Prod.: Michael Harris; Alexander S. Roger
Prod. Agency: Westdeutscher Rundfunk; Outrider International Ltd.; Filmlink Production
Distributor (US): Adler Video Marketing Ltd. (Video sales only)
Traces the changing moral values and social attitudes influencing the photographic representation of the male and female nude in the historical context of photography. Interviews ten photographers working in Europe who explain their own stories and are seen in photo sessions: Jean-François Bauret, Andreas Mahl, Lucien Clergue, Herlinda Koelbl, Kuiston Halle, Franco Fontana, Jeanloup Sieff, Horst P. Horst, Norman Parkinson, and Duane Michals.

REMEMBERING LIFE

57 min. col. 16mm; video 1984
United States English
Dir.: David Hoffman **Prod.:** Alison Abelson
Prod. Agency: Varied Directions
Distributor (US): Varied Directions
Traces the history of *Life* magazine, which constitutes a retrospective of the major events of the twentieth century. Uses documentary material and the reminiscences of photographers and journalists as well as the photojournalism which characterized the magazine. Narrated by news commentator Walter Cronkite.
Reviews: *Booklist,* 15 Jan 86; *Booklist,* 15 Nov 85
Awards: CINE Golden Eagle, 1985

RENCONTRE AVEC JEAN-PIERRE SUDRE ET DENIS BRIHAT

Series Title: Aujourd'hui en France; No. 66
6 min. col. 16mm; video 1980
France French (English, Spanish, and Arabic language versions also avail.)
Prod. Agency: French Ministry of Foreign Affairs
Distributor (US): Embassy of France
Foreign Distr: French Ministry of Foreign Affairs (FR)
Foreign Distr: ICS (FR)

French photographers Jean-Pierre Sudre and Denis Brihat finally accomplish their dream: to leave Paris and live in Provence to be closer to nature. One makes a living by running a photography workshop throughout the school year; the other by holding exhibitions of his work and selling prints. Aujourd'hui en France series, No. 66.

RENCONTRES PHOTOGRAPHIQUES D'ARLES, LES: 1981

Series Title: Aujourd'hui en France; No. 64
5 min. col. 16mm; video 1980
France French (English, Spanish, and Arabic language versions also avail.)
Prod. Agency: French Ministry of Foreign Affairs
Distributor (US): Embassy of France
Foreign Distr: French Ministry of Foreign Affairs (FR)
Foreign Distr: ICS (FR)

Every summer since 1969, the International Photography Festival has organized a unique photographic gathering in Arles, France. The festival serves as a meeting place for amateurs, ever more numerous in France, and also for professionals from all over the world, such as Manuel Alvarez Bravo from Mexico, Linda Wolf from the United States, and Joachim Bonnemaison from France. Aujourd'hui en France series, No. 64.

REPORTERS

97 min. col. 16mm 1981
France French with English subtitles
Dir.: Raymond Depardon
Prod. Agency: Centre Georges Pompidou; Pari Films
Distributor (US): Cinema Guild
Distributor (US): Interama
Foreign Distr: Pari Films (FR)
Foreign Distr: Cinéculture (FR)

French photographer and filmmaker Raymond Depardon (b.1942) looks at the lives of Parisian photojournalists and presents a sympathetic view of the hard-working, unyielding photographers. Depardon followed reporters from the Gamma Agency for the month of October 1980 to record the daily events of the Parisian news scene. Stars some of the leading players in French political and artistic life: François Mitterrand, Jean-Luc Godard, Valéry Giscard d'Estaing, Catherine Deneuve, Jacques Chirac, Alain Delon, Georges Marchais, and Mireille Mathieu. There are guest appearances by visitors to Paris, such as President Bongo of Gabon and American actors Gene Kelly and Richard Gere. (*Cinema Guild*)
Reviews: *Variety; New York Post; New York Times; Arts Weekly; Village Voice*, 21 Oct 86
Awards: César Best Documentary, 1981

RETURN JOURNEY

45 min. col. 16mm; video 1981
Great Britain English
Dir.: Ian Potts **Prod.:** Ian Potts
Prod. Agency: Worktown; Arts Council of Great Britain
Distributor (US): Museum of Modern Art Circ. Film Library
Foreign Distr: Arts Council of Great Britain (GB)

Looks at documentary photography as a way to record working-class life. British artist, photographer, and designer John Humphrey Spender (b.1910) recalls a 1936 anthropological study of Bolton in northern England, where photographs were a useful record of nature and photographers best remained unseen. In 1973, British photographer Derek Smith made the "return journey" to London from the north to study photography. For him, a rapport between subject and photographer is important. Smith questions the intrinsic truthfulness of the documentary photograph in isolation and argues that it is most meaningful when presented in a thematic group, anchored with a text. Compares the work of Spender and Smith with that of Jimmy Forsythe.

RICHARD PRINCE: EDITIONS

7 min. col. video 1982
United States English
Dir.: Carole Ann Klonarides; Michael Owen
Prod.: Carole Ann Klonarides; Michael Owen
Exec. Prod.: Jamie Davidovich
Prod. Agency: MICA-TV
Distributor (US): Electronic Arts Intermix
Distributor (US): Video Data Bank

Using a magazine format similar to that of the "Sixty Minutes" television show, presents the work of American artist and photographer Richard Prince (b.1949), who manipulates magazine advertisements to alter their meaning. Prince narrates experiences which demonstrate his extreme sensitivity to appearances and context. He relates the aesthetic effect that buying his first car had on him.

ROBERT DOISNEAU, BADAUD DE PARIS, PECHEUR D'IMAGES

52 min. col. 16mm; video 1981
France French
Dir.: François Porcile
Prod. Agency: TF 1; Cinéculture
Foreign Distr: Cinéculture (FR)

Profiles French photographer Robert Doisneau (b.1912). Walking with some friends through his favorite places in Paris, Doisneau talks about how the city has changed and the relationship of the photographer to society.

ROBERT FRANK: L'IMMAGINE DELLA POESIA

58 min. col. 16mm 1983
Switzerland Italian
Dir.: Giampiero Tartagni
Prod. Agency: Daria Cinematografica; Radiotelevisione della Svizzera Italiana
Foreign Distr: Radiotelevisione della Svizzera Italiana (SWIT)

Describes the work of Swiss-born American photographer Robert Frank (b.1924). During the 1950s, his still photographs were very influential. In 1958, he began making films, and he went on to teach cinematography. Discusses his films PULL MY DAISY and ME AND MY BROTHER.

ROBERT HEINECKEN

Series Title: On Art and Artists
28 min. b&w video 1976
United States English
Prod.: Kate Horsfield; Lyn Blumenthal
Distributor (US): Video Data Bank

Alex Sweetman interviews American photographer Robert Heinecken (b.1931), who discusses the sexual content of his work, which mingles eroticism and pornography. Talks about his use of images derived from television, mail order catalogues, and advertising. On Art and Artists series.

ROBERT MAPPLETHORPE

Series Title: Arena
60 min. col.& b&w 16mm; video 1988
Great Britain English
Dir.: Nigel Finch
Exec. Prod.: Anthony Wall
Prod. Agency: BBC-TV
Foreign Distr: BBC Enterprises (GB)

Interviews American photographer Robert Mapplethorpe (1946-1989), who discusses portraiture, sexuality, and still lifes. Shows many of his photographs, with commentary by Robin Gibson, curator at the National Gallery in London, art critic Edmund White, and the artist himself. Also interviews portrait subjects, including attorney Michael Stout, art dealers Leo Castelli and Robert Miller, singer Patti Smith, body builder Lisa Lyon, and photographer Lynn Davis. Highlights Mapplethorpe's early experiments in photography and film, with footage of Patti Smith, his companion in the early 1970s. Examines how his photographs of flowers, unconventional homosexual life-styles, and nude studies reveal his consistent attraction to unsentimental, sensual forms. Shows Mapplethorpe at work as he shoots a portrait of American performance artist Laurie Anderson (b.1947) for a record album cover. Arena series.

ROMAN: PHOTO MONTEUR

12 min. col. 35mm 1985
France French (English, Spanish, and Arabic language versions also avail.)
Dir.: Bernard Monsigny
Prod. Agency: Productions Berthemont; Centre National des Arts Plastiques
Distributor (US): Embassy of France
Foreign Distr: Productions Berthemont (FR)
Foreign Distr: ICS (FR)

Photomontage: a form of contemporary plastic art that the Polish graphic artist Roman Cieslewicz (b.1930), who resides in France, has practiced for more than thirty years. He explores images that he removes from their context and then uses to create posters, catalogues, and photographs for advertising.

ROMANTIC HORIZON, THE

Series Title: West of the Imagination, The
58 min. col. video 1986
United States English
Dir.: Tom Weidlinger **Prod.:** David Kennard

Prod. Agency: KERA-TV
Distributor (US): Films for the Humanities
Tells the story of the mythmakers of the American West and the myths they created: in oil or ink, still photographs, and motion picture film. Surveys the historical and cultural landscape of the West. This episode follows the path of explorers Meriwether Lewis (1774-1809) and George Rogers Clark (1752-1818), revealing the lands of the new railroad route to the Pacific through the eyes of artists George Catlin (1796-1872), Karl Bodmer (1809-1893), and Alfred Miller (1810-1874). Narrated by actor James Whitmore. The West of the Imagination series, one of six.
Reviews: *New York Times*, 22 Sept 86; *EFLA Evaluations*, 87; *Sightlines*
Awards: American Film & Video Festival Blue Ribbon, 1987

RUDY BURCKHARDT
Series Title: On Art and Artists
35 min. col. video 1981
United States English
Prod.: Kate Horsfield; Lyn Blumenthal
Distributor (US): Video Data Bank
Robert Storr interviews American photographer and filmmaker Rudy Burckhardt (b.1914), who discusses his work and the differences between still and motion picture photography. On Art and Artists series.

RUE DU REGARD, UN PORTRAIT DE PIERRE GASSMANN
52 min. col. 16mm; video 1988
France French
Dir.: Roger Pic
Prod. Agency: R.I.P.; PICTO
Foreign Distr: Films Grain de Sable (FR)
Profiles French photographer and archivist Pierre Gassmann. After studying at the Bauhaus, he went to Paris, where he became a friend of Brassaï (Gyula Halász), André Kertész, and Man Ray and later archived their negatives. Gassmann has known or handled the works of those who fashioned the look of modernity: surrealists such as Tabard; primitives such as Eugène Atget; László Moholy-Nagy; Emmanuel Sougez; and photojournalists Henri Cartier-Bresson and Robert Capa. In 1952, he established Pictorial Service, an important photographic laboratory in France.

RUTH BERNHARD
Series Title: Photoprofiles
30 min. col. video 1983
United States English
Dir.: John Morgan; Rebecca L. Morgan; Paul R. Schranz; Margaretta K. Mitchell **Prod.:** Thomas R. Schiff
Prod. Agency: Images Productions
Distributor (US): Images Productions
Profiles American photographer Ruth Bernhard (b.1905). Her work, covering more than fifty years, expresses her intense response to the world around her. A respected teacher, Bernhard communicates her perception and philosophy of photography to her students and inspires them to find their own vision. Photoprofiles series.

SAM HASKINS
Series Title: Great Photographers
30 min. col. video 198?
Switzerland English?
Foreign Distr: Polivideo S.A. (SWIT)
Profiles Sam Haskins (b.1926), an advertising photographer from South Africa who now resides in London. Great Photographers series, one of five.

SAM HASKINS
12 min. col. 16mm 1973
Great Britain English
Dir.: W. Webb **Prod.:** R. Marshall
Prod. Agency: London International Film School
Foreign Distr: London International Film School (GB)
Presents South African photographer Sam Haskins (b.1926), now a resident of London, and the creation and subject matter of his photographs.

SCHOENE AUGENBLICK, DER
84 min. col. 16mm 1985
Switzerland German
Dir.: Friedrich Kappeler **Prod.:** Hans-Ulrich Schlumpf
Prod. Agency: Schweizerische Gesellschaft für Volkskunde, Basel
Foreign Distr: Friedrich Kappeler (SWIT)
Foreign Distr: Pio Corradi (SWIT)
Observes three men as they begin to earn their living as street, or itinerant photographers. Examines the varying environments in which they work and their individual expectations.

SECTIONS
5 min. col. 16mm 1981
United States English

Dir.: Sean Phillips
Prod. Agency: University of Southern California, School of Cinema-TV
Distributor (US): University of Southern California Film & Video Distr. Ctr.

Explores streak photography through variations on a walking figure. A streak camera creates an unbroken time record by capturing images without shutter or light interruption.

Awards: CINE Eagle, 1981

SENZENINA: WHAT HAVE WE DONE TO DESERVE THIS?

28 min. col. video 1987
United States English
Dir.: Cato Weatherspoon III **Prod.:** Juanita Anderson; Robert J. Scott
Prod. Agency: WTVS
Distributor (US): WTVS

Presents over two hundred photographs of modern life and struggle in South Africa taken by American photojournalist David C. Turnley, who was on regular assignment there for almost three years.

SHADOW CATCHER, THE

88 min. col.& b&w 16mm; video 1974
United States; Canada English (French language version also avail.)
Dir.: Teri C. McLuhan **Prod.:** Teri C. McLuhan
Distributor (US): Phoenix/BFA Films & Video
Distributor (US): University of Illinois Film Center
Distributor (US): Indiana University Audio Visual Center (Rental only)
Distributor (US): Rental-see *Educational Film & Video Locator*
Foreign Distr: International Tele-Film Enterprises Ltd. (CAN)

Documents the life of American photographer, anthropologist, and filmmaker Edward S. Curtis (1868-1952) and his work with the Native American people from 1896 to 1930; includes all of his recoverable film footage. Excerpts from a film made in 1914 with the Kwakiutl on the Northwest coast show the performance of many masked dances, and there are interviews with three of the original cast in Fort Ruppert, British Columbia. In 1906, Curtis filmed and participated in the Navajo Yebechai Ceremony at the Navajo Indian Reservation in northwestern Arizona and recorded their songs. Another film, made with the Hopi in 1912, appears in the context of a biographical reconstruction of Curtis's initiation into the Snake Fraternity. The sound track includes original Indian music from the Kwakiutl, the Navajo, and the Hopi, as well as several newly composed variations by Comanche composer Eddie Wapp.

Awards: American Film Festival Red Ribbon, 1975

SHADOWS FROM LIGHT

59 min. b/w 16mm; video 1983
Great Britain English
Dir.: Stephen Dwoskin **Prod.:** Robert Jaffé; Kathie Conn
Prod. Agency: Urbane Ltd.; Arts Council of Great Britain
Foreign Distr: Arts Council of Great Britain Film Library (GB)
Foreign Distr: Concord Films Council Ltd. (GB)

A journey through the work of British photographer Bill Brandt (1906-1983), noted for his stark, high-contrast black-and-white compositions. Echoes these compositions with the use of extreme angles, harsh combinations of light and dark, and deep-focus photography, all reminiscent of Orson Welles's CITIZEN KANE. Focuses on Brandt's nudes, his favorite subject.

Awards: Montreal Festival Intl. du Film sur l'Art Best Director, 1984

SHARING TRADITIONS: FIVE BLACK ARTISTS IN NINETEENTH-CENTURY AMERICA

27 min. col. video 1986
United States English
Prod. Agency: Command Communications, Inc.
Distributor (US): Smithsonian Institution

Provides a brief introduction to the life and work of five black American artists of the nineteenth century. Joshua Johnston (c.1765-1830) had a successful business as a portraitist in Baltimore, Maryland. Robert Scott Duncanson (1821-1872) lived in New York and Canada before settling in Cincinnati, Ohio. He painted landscapes, antislavery portraits, and took daguerreotypes. Edward Mitchell Bannister (1828-1901), born in Canada, lived in Providence, Rhode Island, and Boston, Massachusetts, and painted genre scenes of everyday life in New England. Edmonia Lewis (1845-1900), the daughter of a black father and a Chippewa Indian mother, attended Oberlin College in Ohio before traveling to Rome, Italy, where she became famous for her copies of classical sculpture, as well as for figures depicting black and Indian themes. Henry Ossawa Tanner (1859-1937) born in Pittsburgh, Pennsylvania, was the son of a bishop and scholar, and studied in Philadelphia under Thomas Eakins. Tanner's paintings based on religious and biblical themes brought him great success in

Paris, where he moved to escape the racism of his native land. Based on an exhibition catalogue and educational materials prepared by the National Museum of Art/Smithsonian Institution and Chesebrough-Pond's, Inc.

SI J'AVAIS QUATRE DROMADAIRES

52 min. b/w 35mm 1967
France French
Dir.: Chris Marker
Prod. Agency: Iskra
Foreign Distr: Iskra (FR)

Profile of an amateur photographer and his two friends in which they discuss the small fragments of the world captured in their photographs.

SIX FACES OF ROYALTY: VICTORIA

Series Title: Six Faces of Royalty; No. 6
15 min. col. video; 16mm 1973
Great Britain English
Foreign Distr: BBC Enterprises (GB)

The sugary portraits of Queen Victoria by German painter Franz Xaver Winterhalter (1805-1873) and British painter Edwin Henry Landseer (1802-1873) can be contrasted with photographs taken at exactly the same period of her life. By the end of her reign the queen had even been filmed, and there was no disguising the tiny old lady who sat on the throne. Art historian Roy Strong, former director of the National Portrait Gallery in London, demonstrates how we depend on portraits for our ideas about people of the past. Six Faces of Royalty series, No. 6 of 6.

SNOWDON ON CAMERA: PARTS 1 AND 2

Series Title: Snowdon on Camera
2 parts, 40 min. and 45 min. ea. col. video; 16mm 1982
Great Britain English
Prod.: Iain Johnstone
Prod. Agency: BBC Enterprises
Foreign Distr: BBC Enterprises (GB) (Sales only)

British photographer Lord Snowdon (Anthony Charles Robert Armstrong-Jones, b.1930) examines the attitudes and styles of several professional fashion, still-life, and portrait photographers. Snowdon discusses his philosophy with fellow professionals that photography is a simple craft combining daylight and luck. He examines various techniques; talks about two photographers who influenced him; looks at the mechanics and art of photography with famous exponents of the medium; and approaches the valuation of pictures with an open, but slightly skeptical mind. Snowdon on Camera, Parts 1 and 2.

SNOWSCREEN

50 min. col. 16mm; video 1984
Canada English
Dir.: Robert Shoub **Prod.:** Robert Shoub
Distributor (US): American Federation of Arts
Foreign Distr: Robert Shoub (CAN)
Foreign Distr: Canadian Filmmakers Distribution Centre (CAN)

Canadian artist Michael Snow (b.1929) works in a variety of media: painting, photography, music, sculpture, and filmmaking. Looks at examples of his work and includes interviews with Snow and other Canadian artists, dealers, and curators, who discuss his challenging art.
Reviews: *Library Journal*, Vol 3, No 2, 1 Feb 86; *A Newsletter Called FRED...*, Nov-Dec 85

SOCIAL DOCUMENTARY

Series Title: Exploring Photography; No. 4
25 min. col. video; 16mm 1978
Great Britain English
Prod.: Peter Riding
Prod. Agency: BBC Enterprises
Foreign Distr: BBC Enterprises (GB) (Sales only)

Two of Britain's leading photojournalists, Ian Berry (b.1934) and Donald McCullin (b.1935), illustrate the British way of life through strongly contrasting styles. Exploring Photography series, No. 4 of 6.

SOLITUDES

Series Title: Profiles of Nature
25 min. col. 16mm; video 1986
United States English
Dir.: Robert Bocking **Prod.:** Ralph C. Ellis
Prod. Agency: KEG Productions
Distributor (US): Phoenix Films & Video

Observes photographer and naturalist Dan Gibson as he canoes through the Canadian wilderness, recording the sights and sounds of nature. Profiles of Nature series, one of eight.

SOMETHING ABOUT PHOTOGRAPHY

9 min. col. 16mm 1976
United States English
Prod.: Charles Eames; Ray Eames
Distributor (US): Pyramid Film & Video

Shows the importance of the creative side of still photography. Records pointers and insights of American designer and filmmaker Charles Eames (1907-1978) on individual choices and opportunities that one has in the making of each photograph.

SOUTH-EAST NUBA, THE

Series Title: Worlds Apart; No. 2
55 min. col. video; 16mm 1982
Great Britain English
Dir.: Chris Curling; Melissa Llewelyn-Davies
Prod. Agency: BBC-TV
Foreign Distr: BBC Enterprises (GB)

Shows how the publication of photographs of the nude Nuba of Sudan by German filmmaker Leni Riefenstahl (b.1902) led the Sudanese government to bring about changes in the ancient arts and traditions of these people. Worlds Apart series, No. 2.

SOUVENIRS DE JEUNESSE

Series Title: Des Yeux pour Voir
9 min. col. 16mm 1974
France French
Dir.: Michel Pamart
Prod. Agency: ORTF
Foreign Distr: Les Films Constance (FR)
Foreign Distr: INA (FR)

Portrait of French conceptual artist and photographer Christian Boltanski (b.1944). Des Yeux pour Voir series.

STEICHEN: A CENTURY IN PHOTOGRAPHY

60 min. col. video 1980
United States English
Prod. Agency: WXXI-TV
Distributor (US): PBS Video

Profiles American photographer Edward Steichen (1879-1973), whose work encompassed almost all fields of modern photography, including portraiture, advertising, illustration, and combat photography.

STRINDBERG: PEINTRE ET PHOTOGRAPHE

43 min. col. 16mm 1974
Sweden Swedish?
Dir.: Bengt Lagerkvist
Prod. Agency: Swedish Television
Foreign Distr: Centre Culturel de Suède (FR)

Swedish playwright and novelist August Strindberg (1849-1912) took up painting in a free and improvisational style at a time when he was no longer able to write. When he saw he was unable to portray people well enough on canvas, he turned to photography.

STUDIO, THE

Series Title: On Assignment
90 min. col. video 1986
United States English
Dir.: K.C. Lynch **Prod.:** Sally Lynch
Exec. Prod.: Brian D. Ratty
Prod. Agency: Media West Home Video
Distributor (US): Media West Home Video (Video sales only)
Distributor (US): Baker & Taylor Video (Video sales only)

Host photographer Brian D. Ratty offers practical advice on how to lay out and equip a photography studio. Discusses studio design, lighting, classic view cameras and techniques, loading sheet film, shifts, swings, tilts, formal portraits, still lifes, and candid shots. On Assignment series, one of seven.
Reviews: *Shutterbug Magazine*

SUBJECT IS LIGHT, THE

45 min. col. 16mm; video 1986
United States English
Dir.: Richard Wong **Prod.:** Richard Wong
Distributor (US): Acrobat Book Publishers

Presents a series of lighting demonstrations by American photography teacher Charles Potts. Discusses the effects and purposes of light and demonstrates form, texture, and separation by using direct and indirect light. Shows examples of students' photography.

SUSAN HILLER: A PROVISIONAL PORTRAIT

17 min. col. 16mm 1980
Great Britain English
Dir.: Christopher Swayne **Prod.:** Christopher Swayne
Foreign Distr: Arts Council of Great Britain (GB)

Profiles American artist Susan Hiller (b.1940) by recording a week she spent in Matt's Gallery, London, where she unravels a painting thread by thread and ties the threads together in "mental doodles," accelerating the disintegration to defy the myth of the permanence of art. Hiller also seeks to dispel the belief that artists have a special quality or vision by replacing the artist with an ordinary machine in her "photomat portraits" made in automatic photo booths.

SVET ZVANY FESTIVAL

15 min. b&w 35mm 1983
Czechoslovakia Czech (Spanish language version also avail.)

Dir.: Ondrej Balík
Prod. Agency: Krátky Film
Foreign Distr: Czechoslovak Filmexport (CZ)
Shows how a young photographer, Hanka, reacts to the film festival held in Karlovy Var, Czechoslovakia. Passionate about her work and involved in the aesthetic problems of photography, she is disappointed by the festival and finds it a superficial social gathering.

SX-70
11 min. col. 16mm 1972
United States English
Dir.: Charles Eames; Ray Eames
Prod. Agency: Charles and Ray Eames; Polaroid Corp.
Distributor (US): Pyramid Film & Video
Distributor (US): Rental-see *Educational Film & Video Locator*
Describes the SX-70 camera, a system developed by Polaroid that removes the technical barrier between the photographer and the subject.

SYGMA: UNE AGENCE PHOTO: 1980
Series Title: Aujourd'hui en France; No. 42
5 min. col. 16mm; video 1980
France French (English, Spanish, and Arabic language versions also avail.)
Prod. Agency: French Ministry of Foreign Affairs
Distributor (US): Embassy of France
Foreign Distr: French Ministry of Foreign Affairs (FR)
Foreign Distr: ICS (FR)
Forty reporters, three thousand photographs a day, four million negatives on file. Describes how, after eight years, the Sygma Agency constitutes a kind of "world memory bank." Aujourd'hui en France series, No. 42.

SYL LABROT
23 min. b&w video 1977
United States English
Prod. Agency: Video Data Bank
Distributor (US): Video Data Bank
Alex Sweetman interviews American photographer Syl Labrot (1929-1977), who discusses the production of his photography book, *Pleasure Beach.*

TAKING BETTER PICTURES
Series Title: Taking Better Pictures
10 programs, 30 min. ea. col. video 1984
United States English
Prod. Agency: University of Georgia, Center for Education
Distributor (US): Great Plains National Instructional Television Library
Ten-part series designed for beginning and intermediate users of the 35mm camera. The programs are INTRODUCTION; SHUTTER SPEEDS AND APERTURE; INTERCHANGEABLE LENSES; QUALITIES OF LIGHT; EXPOSURE MEASUREMENT; FILM, FILTERS AND SPECIAL EFFECTS; COMPOSITION; NATURE PHOTOGRAPHY; ELECTRONIC FLASH; and PORTRAITS. Taking Better Pictures series.

TECHNIQUE
Series Title: Photography — The Incisive Art
29 min. b&w 16mm; video 1962
United States English
Prod. Agency: KQED-TV; NET
Distributor (US): Indiana University Audio Visual Center
Distributor (US): Pennsylvania State University Audio Visual Services
Distributor (US): International Center of Photography (Archival only)
Distributor (US): Rental-see *Educational Film & Video Locator*
American photographer Ansel Adams (1902-1984) discusses techniques of photography, illustrating with works from his collection the importance of lighting, filters, exposure, magnification, and interpretation in achieving given effects. Photography — The Incisive Art series.

TEMOIN EN QUESTION: LA PHOTOGRAPHIE
50 min. b&w 16mm 1967
France French
Dir.: Charles Prost
Foreign Distr: INA (FR)
Omnipresent in our world, photography is used more and more as a witness to its times. Includes French photographers Brassaï (Gyula Halász), Gabrielle Tessier, Jean-Philippe Charbonnier, Claude Sire, Jean Louis Lerigeon, Jean Louis Clement, Jacques-Henri Lartigue, and Henri Cartier-Bresson and German photographer Gisèle Freund.

TEMPS DE POSES
15 min. col. 16mm 1989
France French
Dir.: Pierre Bressan
Prod. Agency: FR3
Foreign Distr: FR3 (FR)
Visual essay on photography.

TEMPS LIBRES: SPECIAL PHOTO

Series Title: Temps Libres
? min. col. 16mm 1984
France French **Prod.:** Jean Claude Narcy
Foreign Distr: TF 1 (FR)

Magazine-format film on photography. Temps Libres series.

TERRITOIRES PHOTOGRAPHIQUES

Series Title: Territoires Photographiques
8 segments, 12 min. ea. col. video 1985
France French
Dir.: Frédéric Compain; Didier Deleskiewitz; Gilles Delavaux; Olivier Guitton
Prod. Agency: INA; FR3; MCC
Foreign Distr: Cinéculture (FR)

Gives an account of the Mission Photographique of DATAR (Délégation à l'aménagement du territoire et à l'action régionale) and discusses its contribution to the cinematographic representation of the region. Territoires Photographiques series.

THEATRE OF INDIFFERENCE

24 min. col. video 1987
United States English
Dir.: Mitsuru Hayashi
Distributor (US): Facets Multimedia (Video only)

Documents installation projects in New York City by American photographer and sculptor David Finn (b.1921). Visits the East Village, the Bowery, and other Manhattan locations. Contrasts images of local street life with scenes of the artist at work.

THIS IS EDWARD STEICHEN

Series Title: Eye on New York
28 min. b&w 16mm; video 1965
United States English
Dir.: Merrill Brockway **Prod.:** Merrill Brockway
Prod. Agency: WCBS-TV
Distributor (US): Carousel Films
Distributor (US): Indiana University Audio Visual Center (Rental only)
Distributor (US): Rental-see *Educational Film & Video Locator*

American photographer Edward Steichen (1879-1973) elevated portrait photography to an art. Interviewed here at age 86, he discusses his working methods and his philosophy on the relation of nature to art. Steichen recalls his childhood, his stint in Paris, his connection with the 291 Gallery and photographer Alfred Stieglitz, as well as experiences with celebrated friends and photographic subjects, such as sculptor Auguste Rodin and film star Greta Garbo. Offers glimpses of "The Family of Man," his famous 1955 exhibition at the Museum of Modern Art, New York City, and numerous examples of his work. Mrs. Steichen sings a folk song, accompanying herself on guitar, and reads a poem dedicated to Steichen by his brother-in-law Carl Sandburg. Narrated by actor David Wayne. Eye on New York series.
Awards: American Film Festival Blue Ribbon

THREE SHORT LANDSCAPE FILMS

6 min. col. 16mm 1979
Great Britain English
Dir.: Renny Croft
Foreign Distr: London Film-Makers Co-operative (GB)

A three-part film on landscape. The first section explores the relationship between still photography and still film. The second takes the idea of the duplicity of image when related to sound. The third uses a simple change of camera position to reverse the notion of color, space, and perspective.

TIME AND LIGHT

24 min. col. video 1988
Great Britain English
Dir.: Roger Elsgood
Prod. Agency: Arts Council of Great Britain; ILEA Learning Resources Branch
Foreign Distr: Arts Council of Great Britain (GB)

British art historian John Berger (b.1926) presents his critical theory of the nature of photography, using a series of photographs that illustrate the empirical, rhetorical, and aesthetic aspects of the medium. Also explores the relationship between appearance and significance. With the participation of British artist Marc Chaimowicz (b.1947), British photographer Jo Spence (b.1934), and critic Halla Beloff.

TIME MACHINE, THE

56 min. col. 16mm; video 1976
Canada English
Dir.: Terence Macartney-Filgate
Prod.: Terence Macartney-Filgate
Prod. Agency: Canadian Broadcasting Corp.
Foreign Distr: CBC Enterprises (CAN) (Sales only)
Foreign Distr: National Film Board of Canada (CAN)

Presents a history of still photography and discusses its emergence as an art form and as a means of sociopolitical commentary. Beginning with some of the oldest photographs extant, it surveys the legacy of nineteenth-century and twentieth-century faces, cultures, and moods. Features the works and words of eight celebrated photographers: Ansel Adams, Bill Brandt, Brassaï (Gyula Halász), André Kertész, Jacques-Henri Lartigue, Stefan Lorant, Irving Penn, and Paul Strand.

TIME ZERO

11 min. col. 16mm 1980
United States English
Dir.: Glen Fleck **Prod.:** Glen Fleck
Prod. Agency: Glen Fleck Inc.
Distributor (US): Glen Fleck Inc.

Depicts parallels between the interactive process of art and the use of instant photography as a responsive medium.

Awards: CINE Golden Eagle, 1980

TOMORROW IS YOURS: DON MCCULLIN

2 parts, 14 min. and 15 min. col. video 1978
Great Britain English
Prod. Agency: Ryder Cheshire Films
Foreign Distr: Concord Films Council Ltd. (GB)
Foreign Distr: Leonard Cheshire Foundation (GB) (Sales only)

Two-part film in which British photographer Don McCullin (b.1935) talks about some of his experiences of war and major disasters. Shows examples of his photographs from Biafra, Vietnam, and Bangladesh, which illustrate the horrors of war. He also recalls his early life in war-ravaged London, being a misfit at school, and the unlikely way in which his career began.

TOUR DE L'AFRIQUE EN QUATRE-VINGTS JOURS, LE

Series Title: Il Faut le Faire
? min. col. 16mm 1986
France French
Dir.: Jean Michel Boussaguet **Prod.:** ChristianThevenet
Prod. Agency: FR3
Foreign Distr: FR3 (FR)

French photographer Emmanuel Joffet covers the Tour de l'Afrique in eighty days. Il Faut le Faire series.

TRAVELLING LIGHT: THE PHOTOJOURNALISM OF DILIP MEHTA

50 min. col. 16mm 1986
Canada English
Dir.: Deepa Saltzman **Prod.:** Deepa Saltzman
Prod. Agency: Sunrise Films Limited
Foreign Distr: Richard Price Television Associates (GB)

Portrays award-winning photojournalist Dilip Mehta as he travels to Japan, Canada, the United States, England, France, and finally India, where he photographs the commemoration of the anniversary of the Bhopal gas tragedy.

TRIBUTE TO AFRICA

14 min. col. 16mm 1975
United States English
Dir.: Reed Erskine; Susan Erskine **Prod.:** Reed Erskine; Susan Erskine
Prod. Agency: Lightworks Ltd.
Distributor (US): Nat. Mus. of African Art/Smithsonian Inst. (Archival only)

An impressionistic portrait of Africa as seen through the photographs for *Life* magazine by American photographer and Africanist Eliot Elisofon (1911-1973). Contrasts life in big cities and small villages, old and new Africa. Sound track is a mixture of African music and poetry.

Awards: CINE Golden Eagle, 1975

TRUE STORY OF THE CIVIL WAR

33 min. b&w 16mm 1956
United States English
Distributor (US): CRM Films
Distributor (US): Pennsylvania State University Audio Visual Services
Distributor (US): University of Illinois Film Center
Distributor (US): Indiana University Audio Visual Center (Rental only)
Distributor (US): International Center of Photography (Archival only)
Distributor (US): Rental-see *Educational Film & Video Locator*

Actual record made up of the original wet-plate photographs of Mathew B. Brady (1823-1896), newspaper cartoons, and headlines, which are given movement and dramatic impact by special camera and editing techniques. Covers the causes, battles, leaders, and effect of the war upon the nation. Narrated by actor Raymond Massey. (*Penn State*)

TWO PHOTOGRAPHERS: WYNN BULLOCK, IMOGEN CUNNINGHAM

29 min. b&w 16mm; video 1967
United States English
Dir.: Fred Padula
Distributor (US): Canyon Cinema Cooperative
Distributor (US): Rental-see *Educational Film & Video Locator*

A dialogue between two internationally-known American photographers, Wynn Bullock (1902-1972) and Imogen Cunningham (1883-1976), offers insight into their backgrounds, work methods, and personal philosophies. Filmed in their own homes and work spaces, they reveal much of their individual personalities and attitudes.

ULYSSE

22 min. col.& b&w 16mm 1982
France French with English subtitles
Dir.: Agnès Varda
Prod. Agency: Garance
Foreign Distr: Paris Audiovisuel (FR)

Using a photograph of a goat carcass, a man, and a child taken by the filmmaker Agnès Varda in 1954, this autobiographical film-essay evokes the events of the 1950s and presents the people in the photo thirty years later. Interviews the child, Ulysse Lorca, the son of Spanish Republican refugees and now the owner of a bookstore, who cannot remember the circumstances surrounding the photograph. Reflects on the nature of the photographic image and its influence on memory.
Awards: César, 1984

UNCOMMON IMAGES: THE HARLEM OF JAMES VAN DER ZEE

22 min. col.& b&w 16mm; video 1977
United States English
Dir.: Evelyn Barron **Prod.:** Evelyn Barron
Exec. Prod.: Albert Waller
Prod. Agency: WNBC-TV
Distributor (US): Filmakers Library
Distributor (US): Pennsylvania State University Audio Visual Services
Distributor (US): Rental-see *Educational Film & Video Locator*

American photographer James Van Der Zee (1886-1983) reminisces about his life and work, early education, purchase of his first camera at age fourteen, unsuccessful first marriage, and happy second marriage of sixty-five years. Coming to Harlem, New York City, in 1908, he spent the next sixty years compiling a romantic, yet unsentimental record of its middle-class black community: its weddings, funerals, clubs and organizations, church groups, and portraits of the famous and the forgotten.
Reviews: *New York Times*
Awards: CINE Golden Eagle, 1977; Chicago International Film Festival Silver Plaque, 1978; Birmingham International Film Festival, 1979

UNDER THE SEA BY PHOTOSPHERE

Series Title: Travellers in Time — The Trail Blazers
? min. col. video 1986?
Great Britain English
Foreign Distr: BBC Enterprises (GB)

Profiles early-twentieth-century American underwater photographer John Williamson. Includes footage from the first underwater documentary, a 1915 film of divers, corals, wrecks, and sharks, shot in the Bahamas with a four-ton steel photosphere with a big picture window invented by Williamson. Travellers in Time — The Trail Blazers series.

UNQUIET SPIRIT, THE: THE LIFE AND ART OF EDGAR DEGAS (1834-1917)

Series Title: Lively Arts, The
65 min. col. 16mm; video 1980
Great Britain; Federal Republic of Germany English
Dir.: Ann Turner **Prod.:** Ann Turner
Exec. Prod.: Christopher Martin
Prod. Agency: BBC-TV in association with RM Arts; Westdeutscher Rundfunk
Distributor (US): RM Associates/NY
Distributor (US): Films Inc.
Distributor (US): artsAmerica
Distributor (US): Home Vision (Video sales only)
Foreign Distr: BBC Enterprises (GB)
Foreign Distr: Arts Council of Great Britain (GB)
Foreign Distr: RM Associates (GB)

Writer David Thompson profiles French artist Edgar Degas (1834-1917): painter, printmaker, photographer, sculptor, and poet. Describes his evolution from history painter to, Thompson says, "the most penetrating portrayer of modern life of the nineteenth century." Explores the sources of Degas's art: classical traditions, photography, and Japanese prints. Shows many works rarely seen in exhibitions: his notebooks, wax sculptures, and monotypes. The Lively Arts series.
Reviews: *Billboard*, 19 July 86; *Chicago Sun Times*, 1 Aug 86; *Video Insider*, 15 Sept 86; *The News-Herald*, 13 Feb 87; *Los Angeles Times*, 1 Apr 87

URBAN LANDSCAPE
Series Title: Exploring Photography; No. 5
25 min. col. video; 16mm 1978
Great Britain English
Prod.: Peter Riding
Prod. Agency: BBC Enterprises
Foreign Distr: BBC Enterprises (GB) (Sales only)
Shows how photographers attempt to record the environment in which people live. Includes interviews with Chris Killip (b.1946) and American color photographer Joel Meyerowitz (b.1938). Exploring Photography series, No. 5 of 6.

URSULA POWYS-LYBBE
Series Title: Five Women Photographers; No. 4
25 min. col. video 1986
Great Britain English
Dir.: Judy Marle **Prod.:** Ruth Pantoleon
Prod. Agency: Broadside Productions; Channel Four Television
Foreign Distr: Albany Video (GB)
A profile of British photographer Ursula Powys-Lybbe, featuring an interview with her. She discusses her photographs of Australian families in the 1940s. Five Women Photographers, No. 4 of 5.

VALLEE DE LA MORT, LA
Series Title: Camera I
70 min. col.& b&w 16mm 1977
France French
Dir.: Jeanloup Sieff
Foreign Distr: INA (FR)
Using a camera stand and photographs he took of Death Valley, California, French photographer Jeanloup Sieff (b.1933) produced his first feature-length film. It consists of an indirect reflection on photography, a travel journal, and a personal diary. Camera I series.

VICTORIAN PHOTOGRAPHY
36 min. b&w video 1971
Canada English
Dir.: Bill Somerville
Prod. Agency: University of Toronto/Media Centre
Foreign Distr: University of Toronto/Media Centre (CAN) (Video sales only)
Describes improvements in the photographic process during the Victorian era and their impact on both the society and the art of the day. Examines some of the effects of portrait photography on Victorian society and pre-Raphaelite art. Illustrated with photographs from the mid-1800s. Also presents a brief history of the daguerreotype.

VIDEO GUIDE TO BASIC PHOTOGRAPHY, THE
Series Title: On Assignment
90 min. col. video 1988
United States English
Dir.: K.C. Lynch
Exec. Prod.: Brian D. Ratty
Prod. Agency: Media West Home Video
Distributor (US): Media West Home Video (Video sales only)
Distributor (US): Baker & Taylor Video (Video sales only)
Host photographer Brian D. Ratty guides the beginner through basic operations of the Polaroid, 110mm, and 35mm single-lens reflex cameras. Discusses how the cameras work, f-stops and shutter speeds, camera meters, film, available light, flash photography, steady shooting, finding pictures, and portraits. On Assignment series, one of seven.
Reviews: *Shutterbug Magazine*

VIE D'AVENTURES, UNE: FELIX NADAR
52 min. col. 16mm 1979
France French
Dir.: Bernard Miller **Prod.:** Bernard Miller
Foreign Distr: Société Française de Production (FR)
Profiles French photographer Nadar (Gaspard Félix Tournachon, 1820-1910), from his beginnings as a journalist in Paris to his success as a photographer.

VISITE DE L'ATELIER DE MAN RAY, UNE
12 min. col.& b&w 16mm 1981
France French
Dir.: Guy Olivier
Prod. Agency: TF 1
Foreign Distr: TF 1 (FR)
Visits the studio of American painter, photographer, and filmmaker Man Ray (1890-1976). Includes comments by French author and critic Philippe Soupault.

VMCJ
15 min. col. 16mm 1970
United States English
Prod. Agency: United States Navy Department

Distributor (US): National Audiovisual Center
Presents the story of a photographic aerial reconnaissance squadron in combat.

VOYAGE DE ROSE, UN

95 min. col. 35mm 1983
France French
Dir.: Patrick Barbéris
Prod. Agency: Animation Art Graphique Audiovisuel; Association Française pour la Diffusion du Patrimoine Photographique
Foreign Distr: Animation Art Graphique Audiovisuel (FR)
Foreign Distr: Association Française pour la Diffusion du Patrimoine Photographique (FR)
Documents the meeting of French photographer Willy Ronis (b.1910) and Rose Zenner forty-five years after he photographed her during the Citroën strike of 1938. Relates the story of the celebrated picture and describes the role that photography plays in forming people's memories.

W. EUGENE SMITH: PHOTOGRAPHY MADE DIFFICULT

Series Title: American Masters
90 min. col. video 1989
United States English
Dir.: Gene Lasko **Prod.:** Kirk Morris; Marthe Smith
Exec. Prod.: Gregory Andorfer; Susan Lacy
Prod. Agency: WES FORee Productions; WQED-TV; WNET/Channel 13
Distributor (US): Gene Lasko
Distributor (US): American Masters
A docudrama about the life and work of American photographer W. Eugene Smith (1918-1978). Peter Riegert portrays Smith, whose photos captured the changing social conditions and the enduring American spirit both at war and at peace and served as a catalyst for social change. Interviews friends and colleagues, including biographer Jim Hughes; managing editor of *Life* magazine Edward K. Thompson; critic John Berger; photo critic A. D. Coleman; and Cornell Capa of the International Center of Photography, New York. American Masters series.

WALKER EVANS: HIS TIME — HIS PRESENCE — HIS SILENCE

22 min. b&w 16mm 1970
Turkey Nonverbal
Dir.: Sedat Pakay
Distributor (US): Rental-see *Educational Film & Video Locator*
Individual photographic prints by Walker Evans (1903-1975), accompanied by the music of Sidney Bechet and his New Orleans Feetwarmers, Al Jolson and the Andrew Sisters, and the Goldman Band.

WALKING IN A SACRED MANNER

23 min. col. 16mm; video 1982
United States English
Prod.: Stephen Cross
Distributor (US): International Film Bureau
Distributor (US): Rental-see *Educational Film & Video Locator*
Presents photographs of Native Americans by American photographer Edward S. Curtis (1868-1952) made between 1896 and 1930. Combines the photographs with authentic Indian songs and commentary in an attempt to capture the spirit of the North American Indian and the natural world. Includes live footage of the northwestern plains and mountains of the United States.

WAY I SEE IT, THE

28 min. col. 16mm; video 1983
United States English
Dir.: M. Jay Livingston
Prod. Agency: Eastman Kodak Co.
Distributor (US): Baker & Taylor Video (Video sales only)
Distributor (US): Aylmer Press (Video sales only)
Distributor (US): Kodak Video Programs
Distributor (US): University of Illinois Film Center
Portrait photographer Gary Whelpley, nature photographer Janet Paige Bonsall, action photographer Bob Harris, and scenic photographer Norm Kerr relate how they got started, why they selected their specialty, and the tricks of the trade they discovered over the years. Designed to help amateur photographers learn to see what is unique and beautiful and how to record that image.

WEAPONS OF GORDON PARKS, THE

Series Title: Artists at Work
28 min. col. 16mm; video 1966
United States English
Dir.: Warren Forma **Prod.:** Warren Forma
Prod. Agency: Forma Art Associates Ltd.
Distributor (US): Forma Art Associates Ltd.
Distributor (US): International Center of Photography (Archival only)
Distributor (US): Rental-see *Educational Film & Video Locator*
Homage to American photographer, journalist, and film-

maker Gordon Parks (b.1912), based on his photographs and on his autobiography, *A Choice of Weapons*. The fifteenth child of a poor black southern family, Parks overcame the difficulties of his youth in Harlem to become one of *Life* magazine's most noted photographers. A powerful documentation of the impossible living conditions of the poor, his images conveyed a message of love rather than hatred. An exhibition of his photographs concludes the film. (*Films on Art*, 1975 ed.) Artists at Work series.

WEEGEE'S NEW YORK

20 min. col. 16mm 1953
United States English
Distributor (US): Grove Press Films
Distributor (US): International Center of Photography (Archival only)
Distributor (US): Rental-see *Educational Film & Video Locator*

American photographer Weegee (Arthur Fellig, 1900-1968) captures the life and tempo of New York City using documentary and experimental photographic techniques. Contains a humorous sequence photographed candidly at Coney Island which depicts life and love on the beach.

WET PLATE PHOTOGRAPHY PROCESS, THE

4 min. col. video 1983
United States English
Dir.: William Howze
Prod. Agency: Amon Carter Museum
Distributor (US): Amon Carter Museum

Demonstrates the wet-plate photography process through computer animation of nineteenth-century technical manuals. In the 1850s, the wet-plate, or collodion, process revolutionized photography by enabling photographers to make countless prints from glass negatives. Despite the limitations of the cumbersome equipment, early photographers such as Carleton E. Watkins (1829-1916) made photographs of lasting beauty.

WHAT A PICTURE!: FACES AND FIGURES

Series Title: What a Picture!; No. 4
30 min. col. 16mm; video 1983
Great Britain English
Dir.: Chris Haws **Prod.:** Lawrence Moore; Chris Haws
Exec. Prod.: Mick Csaky
Prod. Agency: Mitchell Beazley Ltd.; Thorn-EMI Video
Distributor (US): Films Inc.
Foreign Distr: Thorn-EMI Video (GB) (Video only)

John Hedgecoe, professor of photography at London's Royal College of Art, travels Europe and the United States exploring the world of the professional photographer. He demonstrates how the use of light can flatter the human figure and how to create images of the nude body which are both erotic and romantic. What a Picture! series, No. 4 of 8. (*PMI*)

WHAT A PICTURE!: IMAGINATION AND TECHNIQUE

Series Title: What a Picture!; No. 7
30 min. col. 16mm; video 1983
Great Britain English
Dir.: Chris Haws **Prod.:** Lawrence Moore; Chris Haws
Exec. Prod.: Mick Csaky
Prod. Agency: Mitchell Beazley Ltd.; Thorn-EMI Video
Distributor (US): Films Inc.
Foreign Distr: Thorn-EMI Video (GB) (Video only)

John Hedgecoe, professor of photography at London's Royal College of Art, explores the possibilities of fantasy, imagination, humor, and wit. He explains simple but effective techniques which enable a photographer to conjure up on celluloid almost anything the mind's eye can see. What a Picture! series, No. 7 of 8. (*PMI*)

WHAT A PICTURE!: LANDSCAPE AND LIGHT

Series Title: What a Picture!; No. 5
30 min. col. 16mm; video 1983
Great Britain English
Dir.: Chris Haws **Prod.:** Lawrence Moore; Chris Haws
Exec. Prod.: Mick Csaky
Prod. Agency: Mitchell Beazley Ltd.; Thorn-EMI Video
Distributor (US): Films Inc.
Foreign Distr: Thorn-EMI Video (GB) (Video only)

John Hedgecoe, professor of photography at London's Royal College of Art, travels around Europe capturing mood and atmosphere, investigating viewpoint, composition, and quality of light as he offers insight into the practicalities of landscape photography. What a Picture! series, No. 5 of 8. (*PMI*)

WHAT A PICTURE!: LIGHTING AND COMPOSITION

Series Title: What a Picture!; No. 3
30 min. col. 16mm; video 1983
Great Britain English
Dir.: Chris Haws **Prod.:** Lawrence Moore; Chris Haws
Exec. Prod.: Mick Csaky
Prod. Agency: Mitchell Beazley Ltd.; Thorn-EMI Video

Distributor (US): Films Inc.
Foreign Distr: Thorn-EMI Video (GB) (Video only)

John Hedgecoe, professor of photography at London's Royal College of Art, explains the principles of composition in the selection and arrangement of still-life subjects in a controlled studio situation. Demonstrates the concept of depth of focus in the gardens of a château in Normandy, France. Shows the challenge of photographing a wedding in a picturesque English village church. What a Picture! series, No. 3 of 8. (*PMI*)

WHAT A PICTURE!: MAKING A PICTURE

Series Title: What a Picture!; No. 1
30 min. col. 16mm; video 1983
Great Britain English
Dir.: Chris Haws **Prod.:** Lawrence Moore; Chris Haws
Exec. Prod.: Mick Csaky
Prod. Agency: Mitchell Beazley Ltd.; Thorn-EMI Video
Distributor (US): Films Inc.
Foreign Distr: Thorn-EMI Video (GB)

John Hedgecoe, professor of photography at London's Royal College of Art, introduces the viewer to the principles of the camera and encourages an understanding of the basic elements of light, shape, form, and pattern. The setting of the demonstration is the everyday life of a small family circus. What a Picture! series, No. 1 of 8. (*PMI*)

WHAT A PICTURE!: PERSPECTIVES OF SPACE

Series Title: What a Picture!; No. 6
30 min. col. 16mm; video 1983
Great Britain English
Dir.: Chris Haws **Prod.:** Lawrence Moore; Chris Haws
Exec. Prod.: Mick Csaky
Prod. Agency: Mitchell Beazley Ltd.; Thorn-EMI Video
Distributor (US): Films Inc.
Foreign Distr: Thorn-EMI Video (GB) (Video only)

John Hedgecoe, professor of photography at London's Royal College of Art, visits the châteaux of Normandy, France, and the skyscrapers of New York City to demonstrate architectural photography. Explains the problems of perspective and scale. Looks at the relationship of people to their environments, visiting English fashion designer Zandra Rhodes, the Pearly King and Queen of Westminster in London, and English sculptor Henry Moore (1898-1986). What a Picture! series, No. 6 of 8. (*PMI*)

WHAT A PICTURE!: THE VITAL MOMENT

Series Title: What a Picture!; No. 2
30 min. col. 16mm; video 1983
Great Britain English
Dir.: Chris Haws **Prod.:** Lawrence Moore; Chris Haws
Exec. Prod.: Mick Csaky
Prod. Agency: Mitchell Beazley Ltd.; Thorn-EMI Video
Distributor (US): Films Inc.
Foreign Distr: Thorn-EMI Video (GB) (Video only)

John Hedgecoe, professor of photography at London's Royal College of Art, explains the importance of precision and timing in action photography. Using events such as professional wrestling and white-water canoeing, Hedgecoe extracts lessons in the preparation, anticipation, and perception of the fleeting moment. What a Picture! series, No. 2 of 8. (*PMI*)

WHAT A PICTURE!: TRAVELLING LIGHT

Series Title: What a Picture!; No. 8
30 min. col. 16mm; video 1983
Great Britain English
Dir.: Chris Haws **Prod.:** Lawrence Moore; Chris Haws
Exec. Prod.: Mick Csaky
Prod. Agency: Mitchell Beazley Ltd.; Thorn-EMI Video
Distributor (US): Films Inc.
Foreign Distr: Thorn-EMI Video (GB) (Video only)

John Hedgecoe, professor of photography at London's Royal College of Art, visits Egypt to put into practice the lessons and ideas he has demonstrated in this series of films. What a Picture! series, No. 8 of 8. (*PMI*)

WHITNEY BIENNIAL, THE: PART II

Series Title: ART/new york; No. 7
28 min. col. video 1981
United States English
Dir.: Paul Tschinkel; Marc H. Miller
Prod. Agency: Inner-Tube Video
Distributor (US): Inner-Tube Video

A series on contemporary art that visits New York City galleries and museums and interviews artists and dealers. This segment presents a comprehensive survey of the photography, film, and video shown at the Whitney Museum of American Art's Biennial exhibition in 1981. Includes an interview with video artist Davidson Gigliotti. ART/new york series, No. 7.
Reviews: *Choice*, Jan 83; *Booklist*, Apr 81

WILDLIFE PHOTOGRAPHER AT WORK, A

13 min. col. 16mm 1979
United States English
Prod. Agency: Palisades Pictures; Paramount Communications
Foreign Distr: Educational Foundation for Visual Aids (GB)
Foreign Distr: Fergus Davidson Associates (GB) (Sales only)
Photographer Ken Durgen demonstrates his techniques of photographing wildlife subjects in Jackson Hole, Wyoming.

WILLARD VAN DYKE

Series Title: Photoprofiles
30 min. col. video 1983
United States English
Dir.: John Morgan; Rebecca L. Morgan; Paul R. Schranz; Margaretta K. Mitchell **Prod.:** Thomas R. Schiff
Prod. Agency: Images Productions
Distributor (US): Images Productions
Presents American photographer and filmmaker Willard Van Dyke (1906-1986), a founder of Group f/64, in conversation with his colleague Ralph Steiner. Van Dyke discusses his early values, his interest in Dorothea Lange's photography, and his friendship with Edward Weston as influences on his photographs. Van Dyke also recalls his beginnings in documentary filmmaking. His Gallery 683 in Oakland, California, exhibited works by Weston, Ansel Adams, Imogen Cunningham, and other members of Group f/64. Photoprofiles series.

WILLIAM KLEIN: PAINTER, PHOTOGRAPHER, FILMMAKER

18 min. col. video 1989
Canada English and French with English subtitles
Dir.: Katherine Tweedie
Prod. Agency: Les Productions Extra-Visuelles Inc.
Foreign Distr: Katherine Tweedie (CAN)
Foreign Distr: Cinéma Libre Inc. (CAN)
Profiles American photographer William Klein (b.1926), who has been living and working in Paris since 1949. Describes his provocative methods and the critical responses to his work.

WILLIAM WEGMAN

Series Title: Communications Update
14 min. col. video 1980
United States English
Prod.: Liza Bear; Michael McClard
Distributor (US): Video Data Bank
Profiles American photographer and video artist William Wegman (b.1943). Communications Update series.

WILLY RONIS OU LES CADEAUX DU HASARD

26 min. col. video 1988
France French
Dir.: Patrice Noia
Prod. Agency: Bien Vu; FR3; Centre National des Arts Plastiques; French Ministry of Foreign Affairs; Intermédia; Mission du Patrimoine Photographique; A.F.D.P.P.; Centre National de la Cinématographie
Foreign Distr: Bien Vu (FR)
Profiles French photographer Willy Ronis (b.1910), who composes his images as if they were musical pieces enlivened by the element of chance.

WOMAN BEHIND THE IMAGE, THE: PHOTOGRAPHER JUDY DATER

27 min. col. 16mm 1981
United States English
Dir.: John A. Stewart **Prod.:** John A. Stewart
Prod. Agency: John A. Stewart Productions
Distributor (US): John A. Stewart Productions
Portrait of American photographer Judy Dater (b.1941), revealing the conflict of emotions she experiences as she attempts to keep her work on the edge at all times, while also striving to achieve a fulfilling relationship with a man. The frank presentation of this dilemma, combined with a large sampling of Dater's pictures, helps the viewer to identify the human presence behind a technological art form.
Reviews: *EFLA Evaluations*, Mar-Apr 83

WORLD OF PHOTOGRAPHY, THE

24 min. col. video 1986
United States English
Dir.: William Wegman **Prod.:** William Wegman; Michael Smith
Prod. Agency: The Cat Fund; KTCA
Distributor (US): Electronic Arts Intermix
Distributor (US): Producer Services Group
American postconceptual photographers Michael Smith (b.1942) and William Wegman (b.1943) collaborated on this humorous guide to mastering the techniques, as well as the attitude and approach to subject matter, of photography.

WORLD THAT DISAPPEARED, THE: THE PICTURES OF ROMAN VISHNIAC

30 min. col. video 1978
Switzerland; Great Britain English (German language version also avail.)
Dir.: Erwin Leiser **Prod.:** Erwin Leiser
Prod. Agency: BBC-TV
Foreign Distr: BBC Enterprises (GB)
Foreign Distr: Erwin Leiser Filmproduktion (SWIT)

Profiles American photographer Roman Vishniac (1897-1990), who was born in Russia, educated in Germany, and came to the United States in 1941. He is best known for documenting the lives of Eastern European Jews before World War II. Using a concealed camera, Vishniac photographed the Jewish ghettos of Hungary, Czechoslovakia, and Poland.

WORLD'S GREATEST PHOTOGRAPHY COURSE

88 min. col. video 1983
United States English
Prod. Agency: VidAmerica Inc.
Distributor (US): VidAmerica Inc.
Distributor (US): Baker & Taylor Video (Video sales only)
Distributor (US): University of Illinois Film Center (Video sales only)

Photographer Leif Erikson demonstrates how to take 35mm photographs and discusses the camera, lenses, picture composition, exposure, filters, night photography, and macrophotography.
Reviews: *Booklist*, 1 Nov 85

WYNN BULLOCK: PHOTOGRAPHER

24 min. col. 16mm; video 1977
United States English
Dir.: Thom Tyson **Prod.:** Thom Tyson
Distributor (US): Phoenix/BFA Films & Video
Distributor (US): University of Illinois Film Center
Distributor (US): International Center of Photography (Archival only)
Distributor (US): Rental-see *Educational Film & Video Locator*

Profile of American photographer Wynn Bullock (1902-1972), using footage gathered during the last six months of his life. Interweaves the story of how he and his family faced his terminal cancer with personal reflections on his creative search as a photographer. Underscores Bullock's philosophy of life and art — his reverence for nature and his feeling that light has the power to reveal inner essences — by forsaking the conventional spoken narrative in favor of an evocative musical score featuring Shakuhachi flute master Masayuki Koga. Displays Bullock's photographs by means of a computerized animation technique developed specifically for this production.

ZOLA THE PHOTOGRAPHER (ZOLA PHOTOGRAPHE)

Series Title: Chambre Noire
32 min. b&w 16mm 1969
France French
Dir.: Daniel Georgeot
Foreign Distr: INA (FR)

Profiles French novelist and critic Emile Zola (1840-1902), focusing on his photography. Zola took up photography late in life; nevertheless he considered it not as a pastime but as an art. Photographs from the Zola collection, as well as photographs taken by other amateurs and professionals, recreate the life and times of the novelist. Chambre Noire series.

ZOOM ANGENIEUX, LE

Series Title: Aujourd'hui en France; No. 154
4 min. col. 16mm; video 1985
France French (English, Spanish, and Arabic language versions also avail.)
Prod. Agency: French Ministry of Foreign Affairs
Distributor (US): Embassy of France
Foreign Distr: French Ministry of Foreign Affairs (FR)
Foreign Distr: ICS (FR)

Founded fifty years ago by Pierre Angénieux, the inventor of retrofocus and zoom lenses, the Angénieux Company today is the supplier of lenses for NASA's space missions, as well as for film and television studios all over the world. Aujourd'hui en France series, No. 154.

SUBJECT INDEX

Animals in art
ANIMALS IN MOTION
CHEVAL DE FER, LE
EADWEARD MUYBRIDGE, ZOOPRAXOGRAPHER
MUYBRIDGE IN MOTION
PIONEERS OF PHOTOGRAPHY: THE FLEETING IMAGE

Anthropology — African
SOUTH-EAST NUBA, THE
TRIBUTE TO AFRICA

Anthropology — Native American
SHADOW CATCHER, THE
WALKING IN A SACRED MANNER

Archaeology — Photographic processes and techniques
PHOTOGRAPHIE DEVANT LES BESOINS SPECIFIQUES DES ARCHEOLOGUES, LA

Art and hazardous substances
HEALTH HAZARDS IN PHOTOGRAPHY

Art and literature
FILM SUR HANS BELLMER

Art and politics
JOHN HEARTFIELD: PHOTOMONTAGIST
L.E.A.R.: THE LEAGUE OF REVOLUTIONARY WRITERS AND ARTISTS

Art and revolutions
L.E.A.R.: THE LEAGUE OF REVOLUTIONARY WRITERS AND ARTISTS

Art critics
A.D. COLEMAN
BILL PARKER
ESTELLE JUSSIM
HAL FISCHER READS *AMERICAN PHOTOGRAPHER:* T & A MADE EASY

Art education — Photography
COLLEGE ET L'ALBUM, LE
PHOTOGRAPHER, THE
PHOTOGRAPHIE AMATEUR, LA
PHOTOGRAPHIE, LA: DONNER A VOIR
PHOTOGRAPHIE, LA: TECHNIQUES
PORTRAITS AND DREAMS

Art galleries
PARIS 1985: LA GALERIE DE PHOTOS D'AGATHE GAILLARD

Art market
PHOTOGRAPHIC VISION, THE: THE MARKETPLACE

Artists and models
BERNARD FAUCON
EPOQUE ETAIT DROLE, L'
HARRY CALLAHAN: ELEANOR AND BARBARA
PHOTOSCOPIES
VOYAGE DE ROSE, UN

Asian American artists
EAST MEETS WEST

Black artists
BLACK HAS ALWAYS BEEN BEAUTIFUL
CONVERSATIONS WITH ROY DECARAVA
GORDON PARKS: MOMENTS WITHOUT PROPER NAMES
PHOTOGRAPHE LASSINE, LE
SHARING TRADITIONS: FIVE BLACK ARTISTS IN NINETEENTH-CENTURY AMERICA
UNCOMMON IMAGES: THE HARLEM OF JAMES VAN DER ZEE
WEAPONS OF GORDON PARKS, THE

Children as artists
PORTRAITS AND DREAMS

Civilization — African
SENZENINA: WHAT HAVE WE DONE TO DESERVE THIS?

Civilization — American
IMAGES OF GLORY
PARIS NEW YORK
ROMANTIC HORIZON, THE
TRUE STORY OF THE CIVIL WAR

Civilization — British
SIX FACES OF ROYALTY: VICTORIA

Civilization — French
PARIS NEW YORK
PHOTOGRAPHES ET CRINOLINES

Color theory
COLOR FROM LIGHT
COLOURFUL NOTIONS

Computer art
MUYBRIDGE REVISITED

Conservation — Photographic processes and techniques
KLEINE BLEIBT KLEIN NICHT, DAS

Curators
NATHAN LYONS

Design — Communications design
LIFE AND DEATH OF PICTURE POST, THE
MANDEL/SULTAN
REMEMBERING LIFE
ROMAN: PHOTO MONTEUR

Design — Engineering design
BRIDGEMEN, THE
Erotic art
CORPS PERDU
DAVID HAMILTON
FILM SUR HANS BELLMER
JUST EVERGON
ROBERT HEINECKEN
ROBERT MAPPLETHORPE
Exhibitions
ANDRE KERTESZ OF THE CITIES
BEHIND THE BIG PICTURE
BOLTANSKI
CARLETON WATKINS
DOCUMENTARY URGE, THE: TOM ARNDT
ECRIVAINS PHOTOGRAPHIES PAR ROBERT DOISNEAU, LES
FAMILY OF MAN, THE
FAMILY OF MAN, THE
J.L. SIEFF
KENNETH SNELSON: THE PHOTOGRAPHS
MOSCOU PARIS (1900-1930)
NEW YORK/NEW WAVE AT P.S. 1
PHOTOGRAPHERS OF THE AMERICAN WEST: 1860-1880
PHOTOGRAPHY EXHIBITIONS
SUSAN HILLER: A PROVISIONAL PORTRAIT
WHITNEY BIENNIAL, THE: PART II
Expatriate artists
ALFRED EISENSTAEDT
ANDRE KERTESZ
ANDRE KERTESZ OF THE CITIES
ANDRE KERTESZ, PHOTOGRAPHE AMERICAIN A PARIS: 1981
ANDRE KERTESZ: A POET WITH A CAMERA
AU PAYS DES VISAGES
BANDE A MAN RAY, LA
BRASSAI
CONVERSATION WITH LOTTE JACOBI, A
CORREO CULTURAL
DAVID HOCKNEY'S DIARIES
DREI LEBEN DER ILSE BING, DIE
EISENSTAEDT — GERMANY
EVERYTHING IS PHOTOGRAPH: PROFILE OF ANDRE KERTESZ
FIRE IN THE EAST: A PORTRAIT OF ROBERT FRANK
GISELE FREUND: 1982
HELMUT NEWTON
HOCKNEY THE PHOTOGRAPHER
HORST
LANDSCAPE
LUCIEN AIGNER: A LIFE WITH A CAMERA
MAN RAY
MAN RAY
MAN RAY
MAN RAY PHOTOGRAPHE
MASS AGE IMAGE, THE: PHOTOJOURNALISM; PART I: THE BEGINNINGS OF PHOTOJOURNALISM: ERICH SALOMON; PART II: THE ESSENCE: TIM GIDAL
MOVEMENT
MUSIC IS THE STORY
PHOTOGRAPHERS, THE
PHOTOGRAPHIE ET SOCIETE
ROBERT FRANK: L'IMMAGINE DELLA POESIA
SAM HASKINS
SAM HASKINS
VISITE DE L'ATELIER DE MAN RAY, UNE
WILLIAM KLEIN: PAINTER, PHOTOGRAPHER, FILMMAKER
WORLD THAT DISAPPEARED, THE: THE PICTURES OF ROMAN VISHNIAC
Festivals
ANDRE KERTESZ, PHOTOGRAPHE AMERICAIN A PARIS: 1981
EXPOSITION A NANCY 1909
J'AIME A LA FOLIE...LA PHOTOGRAPHIE
JEAN DIEUZAIDE: 1982
PARIS 1985: LA GALERIE DE PHOTOS D'AGATHE GAILLARD
RENCONTRES PHOTOGRAPHIQUES D'ARLES, LES: 1981
SVET ZVANY FESTIVAL
Film as art
JETEE, LA
NOSTALGIA
Graphic arts
DAVID HOCKNEY'S DIARIES
DEGAS
FILM SUR HANS BELLMER
GEORGE HENDRIK BREITNER
HANS BELLMER
PHOTORAMA
POL MARA
PORTRAITS
ROMAN: PHOTO MONTEUR
UNQUIET SPIRIT, THE: THE LIFE AND ART OF EDGAR DEGAS
Humor and satire
WORLD OF PHOTOGRAPHY, THE
Latin American artists
CORREO CULTURAL
FRIDA KAHLO AND TINA MODOTTI
L.E.A.R.: THE LEAGUE OF REVOLUTIONARY WRITERS AND ARTISTS
MANUEL ALVAREZ BRAVO, FOTOGRAFO
Motion
A. DUMAS OU L'HYPNOSE CINEMA
ANIMALS IN MOTION
CAMERA: EARLY PHOTOGRAPHY: WHEEL OF LIFE — MOVEMENT
CE QUI ME MEUT
CHEVAL DE FER, LE
DEGAS DELACROIX DU REEL A L'IMAGINAIRE
EADWEARD MUYBRIDGE, ZOOPRAXOGRAPHER
FILMING MUYBRIDGE
INTERMITTENCES: E.J. MAREY 1830-1904
MAREY: UN SAVANT TROP MECONNU
MUYBRIDGE IN MOTION
MUYBRIDGE REVISITED
PIONEERS OF PHOTOGRAPHY: THE FLEETING IMAGE
Motion pictures and video
ANNEES-DECLIC, LES
CHEVAL DE FER, LE
COLOR FROM LIGHT
CONVERSATIONS WITH WILLARD VAN DYKE
DAVID BAILEY
DAVID HOWARD: AN ARTIST'S MEDIUM AND METHODS
FIRE IN THE EAST: A PORTRAIT OF ROBERT FRANK
GORDON PARKS: MOMENTS WITHOUT PROPER NAMES
HOLLIS FRAMPTON
MAN RAY
MAN RAY
RALPH STEINER
ROBERT FRANK: L'IMMAGINE DELLA POESIA
RUDY BURCKHARDT
SHADOW CATCHER, THE
SNOWSCREEN
VISITE DE L'ATELIER DE MAN RAY, UNE

WHITNEY BIENNIAL, THE: PART II
WILLARD VAN DYKE
WILLIAM WEGMAN
Motion pictures and video — History
CONNECTIONS: 9-COUNTDOWN
EADWEARD MUYBRIDGE, ZOOPRAXOGRAPHER
LUMIERE PERE ET FILS
MOMENT IN TIME, A
PHOTOGRAPHIE, LA
PIONEERS OF PHOTOGRAPHY: THE FLEETING IMAGE
PREHISTORY OF CINEMA, THE
Multi-media art forms
CONTINUOUS DIARY, THE
JOHN HEARTFIELD: PHOTOMONTAGIST
KENNETH JOSEPHSON
PHOTOMONTAGE TODAY: PETER KENNARD
Mural painting
GEORGES ROUSSE
PORTRAIT D'UN PORTRAITISTE
Museum directors
CORNELL CAPA, PARTS I & II
Museums — Photography museums
FRENCH MUSEUM OF PHOTOGRAPHY, THE
KODAK MUSEUM
PLACE IN THE COUNTRY, A: LACOCK ABBEY, WILTSHIRE
Native American artists
MARTIN CHAMBI AND THE HEIRS OF THE INCAS
Nude in art
ANIMALS IN MOTION
CORPS PERDU
EADWEARD MUYBRIDGE, ZOOPRAXOGRAPHER
HAL FISCHER READS *AMERICAN PHOTOGRAPHER:* T & A MADE EASY
HARRY CALLAHAN: ELEANOR AND BARBARA
JEANLOUP SIEFF
MUYBRIDGE IN MOTION
NEW YORK/NEW WAVE AT P.S. 1
PEINTURES FRAICHES, MADE IN FRANCE
PIONEERS OF PHOTOGRAPHY: THE FLEETING IMAGE
REFLECTIONS
ROBERT MAPPLETHORPE
SHADOWS FROM LIGHT
SOUTH-EAST NUBA, THE
WHAT A PICTURE!: FACES AND FIGURES
WOMAN BEHIND THE IMAGE, THE: PHOTOGRAPHER JUDY DATER
WYNN BULLOCK: PHOTOGRAPHER
Painting — Dada
JOHN HEARTFIELD: PHOTOMONTAGIST
MAN RAY
MAN RAY
MAN RAY
VISITE DE L'ATELIER DE MAN RAY, UNE
Painting — History
SHARING TRADITIONS: FIVE BLACK ARTISTS IN NINETEENTH-CENTURY AMERICA
Painting — Impressionist
DEGAS
DEGAS DELACROIX DU REEL A L'IMAGINAIRE
UNQUIET SPIRIT, THE: THE LIFE AND ART OF EDGAR DEGAS (1834-1917)
Painting — Modernist
CORREO CULTURAL
DU GESTE A L'INFINI: HANS HARTUNG
FILM SUR HANS BELLMER
FRIDA KAHLO AND TINA MODOTTI
HANS BELLMER
JACQUES-HENRI LARTIGUE, PEINTRE ET PHOTOGRAPHE
L.E.A.R.: THE LEAGUE OF REVOLUTIONARY WRITERS AND ARTISTS
MOSCOU PARIS (1900-1930)
PIERRE MICHEL!!! ORCHESTRE A FICTIONS!!!
PORTRAITS
Painting — Portraits
GEORGE HENDRIK BREITNER
PORTRAITS
SIX FACES OF ROYALTY: VICTORIA
Painting — Post-1945
ARNULF RAINER
ARNULF RAINER
ARNULF RAINER
DAVID HOCKNEY'S DIARIES
DAVID HOWARD: AN ARTIST'S MEDIUM AND METHODS
LOUIS JAMMES
PEINTURES FRAICHES, MADE IN FRANCE
PHOTORAMA
SNOWSCREEN
Painting — Post-Impressionist
STRINDBERG: PEINTRE ET PHOTOGRAPHE
Painting — Realist
GEORGE HENDRIK BREITNER
IMAGE MAKERS
IMAGES OF GLORY
ROMANTIC HORIZON, THE
SIX FACES OF ROYALTY: VICTORIA
Painting — Romantic
DEGAS DELACROIX DU REEL A L'IMAGINAIRE
Painting — Surrealist
FILM SUR HANS BELLMER
FRIDA KAHLO AND TINA MODOTTI
HANS BELLMER
MAN RAY
MAN RAY
MAN RAY
VISITE DE L'ATELIER DE MAN RAY, UNE
Photography
TEMPS DE POSES
TEMPS LIBRES: SPECIAL PHOTO
Photography — Aerial views
DIGGING FROM THE AIR
OURSELVES AND THAT PROMISE
PAUL LUTZ
VMCJ
Photography — Aesthetics
A.D. COLEMAN
ART ET LA DECADENCE, L'
ART ET L'ILLUSION, L'
BILL PARKER
CAMERA: EARLY PHOTOGRAPHY: ENDS OF ART
DYNAMO: SPECIAL PHOTO
ESTELLE JUSSIM
HAL FISCHER READS *AMERICAN PHOTOGRAPHER:* T & A MADE EASY
LANGUAGE OF THE CAMERA EYE
MINUTE POUR UNE IMAGE, UNE
MUD OR BLOOD: ISSUES IN PHOTOJOURNALISM
PHOTOGRAPHER'S EYE, THE
PHOTOGRAPHIC VISION, THE: READING PHOTOGRAPHS
PHOTOGRAPHIC VISION, THE: SEEING WITH THE CAMERA
PHOTOGRAPHY
PHOTOGRAPHY AS FINE ART
PHOTOS D'ALIX, LES
PORTRAITS

SNOWDON ON CAMERA: PARTS 1 AND 2
SOMETHING ABOUT PHOTOGRAPHY
SVET ZVANY FESTIVAL
THREE SHORT LANDSCAPE FILMS
TIME AND LIGHT
ULYSSE

Photography — Amateur art
CAMERA: EARLY PHOTOGRAPHY: PHOTOGRAPHIC PLEASURES
CRYSTALS
LEISURE TIME PHOTOGRAPHY
RENCONTRES PHOTOGRAPHIQUES D'ARLES, LES: 1981
SI J'AVAIS QUATRE DROMADAIRES

Photography — Architectural photography
MORLEY BAER
PHOTO TRAVELLING
POINTS OF VIEW
WHAT A PICTURE!: PERSPECTIVES OF SPACE

Photography — Art photography
ART IN AMERICA, PART 9: PHOTOGRAPHY, THE BEGINNING OF A NEW ART
MUD OR BLOOD: ISSUES IN PHOTOJOURNALISM
PHOTOGRAPHY AS FINE ART

Photography — Chromophotography
AUTOCHROMES, LES
COLOURFUL NOTIONS
COULEUR ET PHOTO
GISELE FREUND: 1982
MOVEMENT
PHOTOGRAPHIC VISION, THE: COLOR
PIONEERS OF PHOTOGRAPHY: COLOUR AND THE CAMERA

Photography — City views
ALBEDO
ANDREAS FEININGER
ATGET
BRASSAI
CARLETON WATKINS
CLIMATE OF NEW YORK, THE
DAGUERREOTYPES
EUGENE ATGET
EUGENE ATGET: PHOTOGRAPHER
FIXED IN TIME: A VICTORIAN ALBUM
GLOIRE A FELIX TOURNACHON
IMAGE OF THE CITY, THE
IN THE STREET
JOHN ALBOK'S NEW YORK
JOURNEE AVEC WILLY RONIS, UNE
KENNETH SNELSON: THE PHOTOGRAPHS
OBJECTIF VILLE: ROBERT DOISNEAU
PARIS 09-31
PARIS OF ROBERT DOISNEAU, THE
PHOTOGRAPHIC VISION, THE: LANDSCAPE AND CITYSCAPE
PIONEERS OF PHOTOGRAPHY: NADAR THE GREAT
QUEBEC AS SEEN BY CARTIER-BRESSON, LE
REPORTERS
URBAN LANDSCAPE
WEEGEE'S NEW YORK

Photography — Dance photography
BARBARA MORGAN
BARBARA MORGAN: EVERYTHING IS DANCING

Photography — Fashion photography
BERNARD FAUCON
EPOQUE ETAIT DROLE, L'
FASHION IN THE MAKING
GIANPAOLO BARBIERI
HELMUT NEWTON
HORST
JACQUES-HENRI LARTIGUE, PEINTRE ET PHOTOGRAPHE
LIVES OF LEE MILLER, THE
MODEL
PETER LINDBERGH
PHOTO ET LA MODE, LA

Photography — History
A. DUMAS OU L'HYPNOSE CINEMA
ALFRED STIEGLITZ: PHOTOGRAPHER (1864-1946)
AMERICAN IMAGE: 150 YEARS OF PHOTOGRAPHY
ANDREW J. RUSSELL, A VISUAL HISTORIAN
ANIMALS IN MOTION
ART IN AMERICA, PART 10: PHOTOGRAPHY, MASTERS OF THE TWENTIETH CENTURY
ART IN AMERICA, PART 9: PHOTOGRAPHY, THE BEGINNING OF A NEW ART
ATGET
AUTOCHROMES, LES
CAMERA: EARLY PHOTOGRAPHY: CAUGHT IN TIME
CAMERA: EARLY PHOTOGRAPHY: ENDS OF ART
CAMERA: EARLY PHOTOGRAPHY: EYEWITNESS OF WAR
CAMERA: EARLY PHOTOGRAPHY: FIRST IMPRESSIONS
CAMERA: EARLY PHOTOGRAPHY: MIRROR IMAGE
CAMERA: EARLY PHOTOGRAPHY: NEWSFLASHES
CAMERA: EARLY PHOTOGRAPHY: OUTBACK AND BEYOND
CAMERA: EARLY PHOTOGRAPHY: PHOTOGRAPHIC PLEASURES
CAMERA: EARLY PHOTOGRAPHY: SOLDIERS OF THE QUEEN
CAMERA: EARLY PHOTOGRAPHY: THE OTHER HALF
CAMERA: EARLY PHOTOGRAPHY: TRUTH AND BEAUTY
CAMERA: EARLY PHOTOGRAPHY: WHEEL OF LIFE — MOVEMENT
CAMERA: EARLY PHOTOGRAPHY: WORLD ON A PLATE
CE QUI ME MEUT
CHEVAL DE FER, LE
CONNECTIONS: 9-COUNTDOWN
DAGUERRE: THE BIRTH OF PHOTOGRAPHY
DEGAS DELACROIX DU REEL A L'IMAGINAIRE
EADWEARD MUYBRIDGE, ZOOPRAXOGRAPHER
EARLY PHOTOGRAPHERS, THE
ERNEST LIVERNOIS, PHOTOGRAPHE
EUGENE ATGET
EUGENE ATGET: PHOTOGRAPHER
FILMING MUYBRIDGE
FIXED IN TIME: A VICTORIAN ALBUM
FOCUS ON PHOTOGRAPHY
FRENCH MUSEUM OF PHOTOGRAPHY, THE
FRUEHE PHOTOGRAPHISCHE BILDNISSE IM KUPFERSTICH-KABINETT
GEORGE EASTMAN: ONE MAN'S VISION, IMAGE FOR ALL
GLOIRE A FELIX TOURNACHON
HERITAGE FABULEUX DE NICEPHORE NIEPCE, L'
IMAGE MAKERS
IMAGES
IMAGES OF GLORY
IN CAMERA
INSTANT POUR L'ETERNITE, UN

INTERMITTENCES: E.J. MAREY 1830-1904
INVENTION DE LA PHOTOGRAPHIE, L'
INVENTION DE LA PHOTOGRAPHIE, L'
INVENTORS, THE
KEEPERS OF THE LIGHT
LANDSCAPE PHOTOGRAPHERS
LEONARD MISONNE: THE PHOTOGRAPHER
LITTLE BIT OF MAGIC REALISED, A
LUMIERE PERE ET FILS
MAREY: UN SAVANT TROP MECONNU
MARINE TERRACE
MATHEW BRADY: PHOTOGRAPHER OF AN ERA
MIROIR DE PAPIER, LE
MOMENT IN TIME, A
MOVING STILL
MUYBRIDGE IN MOTION
MUYBRIDGE REVISITED
NADAR
NAKED EYE, THE
OEIL NEUF, UN
OTHER OBSERVERS, THE: WOMEN PHOTOGRAPHERS FROM 1850 TO THE PRESENT DAY
PHOTO ET LA MODE, LA
PHOTOGRAPHERS OF THE AMERICAN WEST: 1860-1880
PHOTOGRAPHES ET CRINOLINES
PHOTOGRAPHIC VISION, THE: A VISUAL HERITAGE
PHOTOGRAPHIC VISION, THE: VISUAL DOCUMENTS
PHOTOGRAPHIE, LA
PHOTOGRAPHY
PIONEERS OF PHOTOGRAPHY: COLOUR AND THE CAMERA
PIONEERS OF PHOTOGRAPHY: FAMOUS MEN AND FAIR WOMEN
PIONEERS OF PHOTOGRAPHY: MIRROR WITH A MEMORY
PIONEERS OF PHOTOGRAPHY: NADAR THE GREAT
PIONEERS OF PHOTOGRAPHY: SUN PICTURES
PIONEERS OF PHOTOGRAPHY: THE FLEETING IMAGE
PIONEERS OF PHOTOGRAPHY: THE PENCIL OF NATURE
PIONEERS OF PHOTOGRAPHY: TRAVELLING MAN
PIONNIERS, LES
PREHISTORY OF CINEMA, THE
PRIMITIFS DE LA PHOTOGRAPHIE, LES
QUAI BOURBON
QUELLE BELLE EPOQUE
REALISME ET ILLUSION
REFLECTIONS
ROMANTIC HORIZON, THE
SHADOW CATCHER, THE
SHARING TRADITIONS: FIVE BLACK ARTISTS IN NINETEENTH-CENTURY AMERICA
SIX FACES OF ROYALTY: VICTORIA
STRINDBERG: PEINTRE ET PHOTOGRAPHE
TIME MACHINE, THE
TRUE STORY OF THE CIVIL WAR
UNQUIET SPIRIT, THE: THE LIFE AND ART OF EDGAR DEGAS (1834-1917)
VICTORIAN PHOTOGRAPHY
VIE D'AVENTURES, UNE: FELIX NADAR
WALKING IN A SACRED MANNER
WET PLATE PHOTOGRAPHY PROCESS, THE
ZOLA THE PHOTOGRAPHER

Photography — Instant camera photographs
ART OF MARIE COSINDAS, THE
BEHIND THE BIG PICTURE
COLOURFUL NOTIONS
DAVID HOCKNEY'S DIARIES
JUST EVERGON
SX-70
TIME ZERO

Photography — Landscapes
ANSEL ADAMS
ANSEL ADAMS: PHOTOGRAPHER
ANSEL ADAMS: PHOTOGRAPHER
ANSEL ADAMS: PHOTOGRAPHER OF THE AMERICAN WEST
BARBARA MORGAN
CARLETON WATKINS
ELIOT PORTER
ELIOT PORTER'S WORLD
FAY GODWIN
GREAT DAYS IN THE ROCKIES
LANDSCAPE
LANDSCAPE PHOTOGRAPHER
LANDSCAPE PHOTOGRAPHERS
MORLEY BAER
OURSELVES AND THAT PROMISE
PHOTOGRAPHERS OF EVERYDAY LIFE
PHOTOGRAPHERS OF THE AMERICAN WEST: 1860-1880
PHOTOGRAPHIC VISION, THE: LANDSCAPE AND CITYSCAPE
PHOTOGRAPHY AS AN ART
VALLEE DE LA MORT, LA
WHAT A PICTURE!: LANDSCAPE AND LIGHT

Photography — Nature photography
BEYOND THE OCEAN: BENEATH A LEAF
CRYSTALS
ELIOT PORTER
L.L. BEAN GUIDE TO OUTDOOR PHOTOGRAPHY
LENSES ON NATURE
NEVER GIVE UP — IMOGEN CUNNINGHAM
SOLITUDES
WILDLIFE PHOTOGRAPHER AT WORK, A

Photography — Panoramas
KENNETH SNELSON: THE PHOTOGRAPHS

Photography — Photographers
AARON SISKIND
ALBERT TESSIER A FORCE D'IMAGES
ALFRED EISENSTAEDT
ALFRED STIEGLITZ: PHOTOGRAPHER (1864-1946)
AMERICA AND LEWIS HINE
ANDRE KERTESZ
ANDRE KERTESZ OF THE CITIES
ANDRE KERTESZ: A POET WITH A CAMERA
ANDREAS FEININGER
ANDREAS FEININGER
ANNEES-DECLIC, LES
ANSEL ADAMS
ANSEL ADAMS: PHOTOGRAPHER
ANSEL ADAMS: PHOTOGRAPHER
ANSEL ADAMS: PHOTOGRAPHER OF THE AMERICAN WEST
ARNULF RAINER
ARNULF RAINER
ARNULF RAINER
ART OF MARIE COSINDAS, THE
ATHLETE, THE: ROBERT RIGER'S DRAWINGS, PHOTOGRAPHS AND TELEVISION SPORTS
BARBARA MORGAN
BARBARA MORGAN: EVERYTHING

IS DANCING
BENNO FRIEDMAN
BERNARD FAUCON
BERNARD FAUCON, PHOTOGRAPHE
BERT HARDY
BETTY HAHN
BILL BRANDT
BOLTANSKI
BRASSAI
CARLETON WATKINS
CE QUI ME MEUT
CHINE, VUES INTERIEURES
CONTACTS
CONTEMPORARY PHOTOGRAPHERS: AL WEBER
CONTEMPORARY PHOTOGRAPHERS: DAVID CAVAGNARD
CONTEMPORARY PHOTOGRAPHERS: MORLEY BAER
CONTEMPORARY PHOTOGRAPHERS: RUTH BERNHARD
CONTEMPORARY PHOTOGRAPHERS: STEVE CROUCH
CONVERSATION WITH LOTTE JACOBI, A
CONVERSATIONS WITH ROY DECARAVA
CONVERSATIONS WITH WILLARD VAN DYKE
CORNELL CAPA, PARTS I & II
DAVID BAILEY
DAVID GOLDBLATT IN BLACK AND WHITE
DAVID HAMILTON
DAVID HARTMAN: THE SHOOTERS
DAYBOOKS OF EDWARD WESTON, THE: HOW YOUNG I WAS
DAYBOOKS OF EDWARD WESTON, THE: THE STRONGEST WAY OF SEEING
DOCUMENTARILY THINE: LEWIS HINE
DOROTHEA LANGE
DOROTHEA LANGE: THE CLOSER FOR ME
DOROTHEA LANGE: UNDER THE TREES
DREI LEBEN DER ILSE BING, DIE
DU GESTE A L'INFINI: HANS HARTUNG
DUANE MICHALS (1939-1997)
DYNAMO: SPECIAL PHOTO
EAST MEETS WEST
EDWARD STEICHEN
EDWARD STEICHEN
EDWARD WESTON: ARTIST OF THE CAMERA
EISENSTAEDT — GERMANY
ELIOT PORTER
ELIOT PORTER'S WORLD
ENFANCE D'UN PHOTOGRAPHE: LARTIGUE
EVERYTHING IS PHOTOGRAPH: PROFILE OF ANDRE KERTESZ
EVON STREETMAN
EXTRAITS DU JOURNAL DE JACQUES-HENRI LARTIGUE
EYE OF CARTIER-BRESSON, THE
FACE TO FACE
FAY GODWIN
FIRE IN THE EAST: A PORTRAIT OF ROBERT FRANK
FRIDA KAHLO AND TINA MODOTTI
GEORGE HENDRIK BREITNER
GEORGE TICE
GIANPAOLO BARBIERI
GOING WHERE I'VE NEVER BEEN: PHOTOGRAPHY OF DIANE ARBUS
GORDON PARKS: MOMENTS WITHOUT PROPER NAMES
GRACE ROBERTSON
GUY LE QUERREC, PHOTOGRAPHE
HANS BELLMER
HANS NAMUTH, INSTANTS DONNES
HANS STAUB, PHOTOJOURNALIST
HARRY CALLAHAN: A BROADCAST EXHIBITION
HARRY CALLAHAN: A NEED TO SEE AND EXPRESS
HARTUNG
HELEN MUSPRATT
HELMUT NEWTON
HOCKNEY THE PHOTOGRAPHER
HOLLIS FRAMPTON
HOMAGE TO AUGUST SANDER
HORST
IMAGE OF LIGHT: THE PHOTOGRAPHY OF GEORGE POLLOCK
IMOGEN CUNNINGHAM AT 93
IMOGEN CUNNINGHAM, PHOTOGRAPHER
JACQUES-HENRI LARTIGUE
JACQUES-HENRI LARTIGUE
JACQUES-HENRI LARTIGUE, PEINTRE ET PHOTOGRAPHE
JACQUES-HENRI LARTIGUE: 1980
JEAN DIEUZAIDE: 1982
JEANLOUP SIEFF
JOHN ALBOK'S NEW YORK
JOHN BALDESSARI
JOHN BALDESSARI: A CONVERSATION
JOHN HOAGLAND: FRONTLINE PHOTOGRAPHER
JUST EVERGON
KARSH: THE SEARCHING EYE
KEITH SMITH
KENNETH SNELSON
LAURA GILPIN: AN ENDURING GRACE
LEONARD MISONNE: THE PHOTOGRAPHER
LIVES OF LEE MILLER, THE
LOUIS JAMMES
LUCIEN AIGNER: A LIFE WITH A CAMERA
LUMIERE PERE ET FILS
MAN RAY
MAN RAY
MAN RAY
MAN RAY PHOTOGRAPHE
MANDEL/SULTAN
MANUEL ALVAREZ BRAVO, FOTOGRAFO
MARCIA RESNICK
MAREY: UN SAVANT TROP MECONNU
MARGARET BOURKE-WHITE
MARGARET MONCK
MARTIN CHAMBI AND THE HEIRS OF THE INCAS
MILTON ROGOVIN
MILTON ROGOVIN
MORLEY BAER
NATHAN LYONS
NEAL SLAVIN
NEVER GIVE UP — IMOGEN CUNNINGHAM
OLIVER GAGLIANI
OTHER OBSERVERS, THE: WOMEN PHOTOGRAPHERS FROM 1850 TO THE PRESENT DAY
PATRICK ULLMANN, PHOTOGRAPHE DES CHANTEURS FRANCAIS
PAUL LUTZ
PHILIP TRAGER
PHOTOGRAPHE WILLY RONIS, LE
PHOTOGRAPHER, THE
PHOTOGRAPHER, THE
PHOTOGRAPHER: RUSSELL LEE

PHOTOGRAPHERS OF EVERYDAY LIFE
PHOTOGRAPHERS, THE
PHOTOPROFILES
PHOTORAMA
PIERRE MICHEL!!! ORCHESTRE A FICTIONS!!!
PIONEERS OF PHOTOGRAPHY: FAMOUS MEN AND FAIR WOMEN
PIONEERS OF PHOTOGRAPHY: NADAR THE GREAT
PIONEERS OF PHOTOGRAPHY: SUN PICTURES
PIONEERS OF PHOTOGRAPHY: THE PENCIL OF NATURE
PIONEERS OF PHOTOGRAPHY: TRAVELLING MAN
PISTA: THE MANY FACES OF STEPHEN DEUTCH
POL MARA
PORTFOLIO
PORTRAIT DE JEAN DIEUZAIDE
PORTRAIT OF IMOGEN
PORTRAITS
PORTRAITS ET GRIMACES: WITKIEWICZ
RALPH STEINER
REAL WEEGEE, THE
RENCONTRE AVEC JEAN-PIERRE SUDRE ET DENIS BRIHAT
RICHARD PRINCE: EDITIONS
ROBERT DOISNEAU, BADAUD DE PARIS, PECHEUR D'IMAGES
ROBERT FRANK: L'IMMAGINE DELLA POESIA
ROBERT HEINECKEN
ROBERT MAPPLETHORPE
RUDY BURCKHARDT
RUE DU REGARD, UN PORTRAIT DE PIERRE GASSMANN
RUTH BERNHARD
SAM HASKINS
SAM HASKINS
SCHOENE AUGENBLICK, DER
SHADOW CATCHER, THE
SHADOWS FROM LIGHT
SNOWSCREEN
SOUVENIRS DE JEUNESSE
STEICHEN: A CENTURY IN PHOTOGRAPHY
STRINDBERG: PEINTRE ET PHOTOGRAPHE
SYL LABROT
THIS IS EDWARD STEICHEN
TOMORROW IS YOURS: DON MCCULLIN
TWO PHOTOGRAPHERS: WYNN BULLOCK, IMOGEN CUNNINGHAM
UNCOMMON IMAGES: THE HARLEM OF JAMES VAN DER ZEE
URSULA POWYS-LYBBE
VISITE DE L'ATELIER DE MAN RAY, UNE
W. EUGENE SMITH: PHOTOGRAPHY MADE DIFFICULT
WEAPONS OF GORDON PARKS, THE
WILLARD VAN DYKE
WILLIAM KLEIN: PAINTER, PHOTOGRAPHER, FILMMAKER
WILLIAM WEGMAN
WILLY RONIS OU LES CADEAUX DU HASARD
WOMAN BEHIND THE IMAGE, THE: PHOTOGRAPHER JUDY DATER
WORLD THAT DISAPPEARED, THE: THE PICTURES OF ROMAN VISHNIAC
WYNN BULLOCK: PHOTOGRAPHER
ZOLA THE PHOTOGRAPHER

Photography — Photographic tools and equipment

ACTION PHOTOGRAPHY
APPAREIL PHOTOGRAPHIQUE, L'
DARKROOM, THE
HEALTH HAZARDS IN PHOTOGRAPHY
LICHFIELD ON PHOTOGRAPHY
PHOTO SHOW, THE
PHOTOGRAPHIC DESIGN
PHOTOGRAPHIC LIGHT
PHOTOGRAPHIC VISION, THE: THE CAMERA
PHOTOGRAPHIC VISION, THE: THE FILM
STUDIO, THE
VIDEO GUIDE TO BASIC PHOTOGRAPHY, THE
ZOOM ANGENIEUX, LE

Photography — Photojournalism

ALFRED EISENSTAEDT
AMERICA AND LEWIS HINE
AMERICAN IMAGES: PHOTOGRAPHS AND PHOTOGRAPHERS OF THE FARM SECURITY ADMINISTRATION, 1932-1942
ANNEES KAGAN, LES
ANNEES-DECLIC, LES
BERT HARDY
BRIDGEMEN, THE
CAMERA: EARLY PHOTOGRAPHY: NEWSFLASHES
CIVIL WAR PHOTOGRAPHERS: WITNESS TO THE STORM
CORNELL CAPA, PARTS I & II
DAVID GOLDBLATT IN BLACK AND WHITE
DOCUMENTARILY THINE: LEWIS HINE
DOCUMENTARY URGE, THE: TOM ARNDT
DOROTHEA LANGE
DOROTHEA LANGE: THE CLOSER FOR ME
DOROTHEA LANGE: UNDER THE TREES
EISENSTAEDT — GERMANY
ENTREE LIBRE
FIRE IN THE EAST: A PORTRAIT OF ROBERT FRANK
FLAGRANTS DELITS
GORDON PARKS: MOMENTS WITHOUT PROPER NAMES
GRACE ROBERTSON
GRAND REPORTER
HANS STAUB, PHOTOJOURNALIST
JOHN HOAGLAND: FRONTLINE PHOTOGRAPHER
"LET US NOW PRAISE FAMOUS MEN": ALABAMA 40 YEARS ON
LIFE AND DEATH OF PICTURE POST, THE
LIVES OF LEE MILLER, THE
MARGARET BOURKE-WHITE
MASS AGE IMAGE, THE: PHOTOJOURNALISM; PART I: THE BEGINNINGS OF PHOTOJOURNALISM: ERICH SALOMON; PART II: THE ESSENCE: TIM GIDAL
MATHEW BRADY: PHOTOGRAPHER OF AN ERA
MOVING STILLS
MUD OR BLOOD: ISSUES IN PHOTOJOURNALISM
ONE SEEING...
PHOTO JOURNALISME
PHOTOGRAPHER: RUSSELL LEE
PHOTOGRAPHERS OF EVERYDAY LIFE
PHOTOGRAPHIC VISION, THE: PHOTOJOURNALISM
PHOTOGRAPHIE ET SOCIETE
PHOTOGRAPHY

PHOTOJOURNALISM
PICTURING DERRY
QUEBEC AS SEEN BY CARTIER-BRESSON, LE
RAPHO, HISTOIRE D'UNE FAMILLE
REAL WEEGEE, THE
RECITS D'ELLIS ISLAND: TRACES
REMEMBERING LIFE
REPORTERS
RETURN JOURNEY
ROBERT DOISNEAU, BADAUD DE PARIS, PECHEUR D'IMAGES
ROBERT FRANK: L'IMMAGINE DELLA POESIA
SENZENINA: WHAT HAVE WE DONE TO DESERVE THIS?
SOCIAL DOCUMENTARY
SOUTH-EAST NUBA, THE
SYGMA: UNE AGENCE PHOTO: 1980
TEMOIN EN QUESTION: LA PHOTOGRAPHIE
TOMORROW IS YOURS: DON MCCULLIN
TOUR DE L'AFRIQUE EN QUATRE-VINGTS JOURS, LE
TRAVELLING LIGHT: THE PHOTO JOURNALISM OF DILIP MEHTA
TRIBUTE TO AFRICA
TRUE STORY OF THE CIVIL WAR
VOYAGE DE ROSE, UN
W. EUGENE SMITH: PHOTOGRAPHY MADE DIFFICULT
WALKER EVANS: HIS TIME — HIS PRESENCE — HIS SILENCE
WEAPONS OF GORDON PARKS, THE

Photography — Photomontage
CONTINUOUS DIARY, THE
JOHN HEARTFIELD: PHOTO-MONTAGIST
PEINTURES FRAICHES, MADE IN FRANCE
PHOTOMONTAGE TODAY: PETER KENNARD
ROMAN: PHOTO MONTEUR

Photography — Portraits
ALFRED EISENSTAEDT
ANDERSEN CHEZ LE PHOTOGRAPHE
ANSEL ADAMS: PHOTOGRAPHER
AU PAYS DES VISAGES
BANDE A MAN RAY, LA
BARBARA MORGAN
BLACK HAS ALWAYS BEEN BEAUTIFUL
CAMERA: EARLY PHOTOGRAPHY: MIRROR IMAGE
CINDY SHERMAN: AN INTERVIEW
CONVERSATION WITH LOTTE JACOBI, A
DAVID GREMP: JESSICA, JULY 15, 1983
ECRIVAINS PHOTOGRAPHIES PAR ROBERT DOISNEAU, LES
EISENSTAEDT — GERMANY
FACE TO FACE
FAMILY OF MAN, THE
FAMILY OF MAN, THE
FRUEHE PHOTOGRAPHISCHE BILDNISSE IM KUPFERSTICH-KABINETT
GEORGE HENDRIK BREITNER
GISELE FREUND: 1982
GLOIRE A FELIX TOURNACHON
GOING WHERE I'VE NEVER BEEN: PHOTOGRAPHY OF DIANE ARBUS
HANS NAMUTH, INSTANTS DONNES
HARRY CALLAHAN: ELEANOR AND BARBARA
HELMUT NEWTON
JACQUES-HENRI LARTIGUE, PEINTRE ET PHOTOGRAPHE
JEANLOUP SIEFF
KARSH: THE SEARCHING EYE
LOCAL STRANGER, A
MARGINA 88
MARTIN CHAMBI AND THE HEIRS OF THE INCAS
NEVER GIVE UP — IMOGEN CUNNINGHAM
NEW YORK/NEW WAVE AT P.S. 1
PATRICK ULLMANN, PHOTOGRAPHE DES CHANTEURS FRANCAIS
PHOTOGRAPHE LASSINE, LE
PHOTOGRAPHIC VISION, THE: THE PORTRAIT
PHOTOGRAPHIE ET SOCIETE
PHOTOSCOPIES
PIONEERS OF PHOTOGRAPHY: FAMOUS MEN AND FAIR WOMEN
PIONEERS OF PHOTOGRAPHY: NADAR THE GREAT
POINTS OF VIEW
PORTRAIT D'UN PORTRAITISTE
PORTRAIT, THE
PORTRAITS
PORTRAITS ET GRIMACES: WITKIEWICZ
ROBERT MAPPLETHORPE
SIX FACES OF ROYALTY: VICTORIA
SOUTH-EAST NUBA, THE
SUSAN HILLER: A PROVISIONAL PORTRAIT
TEMOIN EN QUESTION: LA PHOTOGRAPHIE
THIS IS EDWARD STEICHEN
ULYSSE
UNCOMMON IMAGES: THE HARLEM OF JAMES VAN DER ZEE
WOMAN BEHIND THE IMAGE, THE: PHOTOGRAPHER JUDY DATER

Photography — Postcards
CARTES POSTALES VIDEO
CORPS PERDU
QUELLE BELLE EPOQUE

Photography — Sports photography
ACTION PHOTOGRAPHY
ATHLETE, THE: ROBERT RIGER'S DRAWINGS, PHOTOGRAPHS AND TELEVISION SPORTS

Photography — Technique
ACTION PHOTOGRAPHY
AUTOUR DE LA PHOTOGRAPHIE
BANLIEUES ENCEINTES
BENNO FRIEDMAN
BETTY HAHN
CARLOTTA CORPRON: DESIGNER WITH LIGHT
CIRCLE OF LIGHT: THE PHOTO-GRAPHY OF PAMELA BONE
COLOR FROM LIGHT
DARKROOM, THE
DAVID GREMP: JESSICA, JULY 15, 1983
DAVID HOWARD: AN ARTIST'S MEDIUM AND METHODS
DERNIER ADIEU, DE JEAN MARC TINGAUD
FREEMAN IN TIME
FROM THE LOG OF JUST LOVE
HERITAGE FABULEUX DE NICEPHORE NIEPCE, L'
IMAGE, THE
JE TE PRENDS, TU ME PRENDS... EN PHOTO
KENNETH JOSEPHSON
KLEINE BLEIBT KLEIN NICHT, DAS
KODAK'S HOW TO TAKE BETTER PICTURES
L.L. BEAN GUIDE TO OUTDOOR PHOTOGRAPHY
LEISURE TIME PHOTOGRAPHY

LICHFIELD ON PHOTOGRAPHY
MASTER OF THE CAMERA: EDWARD J. STEICHEN
MOVEMENT
PHOTO SHOW, THE
PHOTO TIPS
PHOTO TRAVELLING
PHOTOGRAPHIC DESIGN
PHOTOGRAPHIC LIGHT
PHOTOGRAPHIC VISION, THE: A LIFETIME IN PHOTOGRAPHY
PHOTOGRAPHIC VISION, THE: ARTISTIC EXPRESSION
PHOTOGRAPHIC VISION, THE: COLOR
PHOTOGRAPHIC VISION, THE: CONTROLLING EXPOSURE
PHOTOGRAPHIC VISION, THE: IMAGE AND IMPACT
PHOTOGRAPHIC VISION, THE: LANDSCAPE AND CITYSCAPE
PHOTOGRAPHIC VISION, THE: PHOTOJOURNALISM
PHOTOGRAPHIC VISION, THE: READING PHOTOGRAPHS
PHOTOGRAPHIC VISION, THE: RESPONDING TO LIGHT
PHOTOGRAPHIC VISION, THE: SEEING WITH THE CAMERA
PHOTOGRAPHIC VISION, THE: STRUCTURE WITHIN THE IMAGE
PHOTOGRAPHIC VISION, THE: THE CAMERA
PHOTOGRAPHIC VISION, THE: THE DARKROOM
PHOTOGRAPHIC VISION, THE: THE FILM
PHOTOGRAPHIC VISION, THE: THE PORTRAIT
PHOTOGRAPHIC VISION, THE: THE STUDIO
PHOTOGRAPHIC VISION, THE: TIME AND MOTION
PHOTOGRAPHY AS AN ART
PICTURES
PROCEDE FRESSON, LE
PROFESSIONAL PHOTOGRAPHY
SECTIONS
SNOWDON ON CAMERA: PARTS 1 AND 2
SOMETHING ABOUT PHOTOGRAPHY
STUDIO, THE
SUBJECT IS LIGHT, THE
SX-70
TAKING BETTER PICTURES
TECHNIQUE
VIDEO GUIDE TO BASIC PHOTOGRAPHY, THE
WAY I SEE IT, THE
WET PLATE PHOTOGRAPHY PROCESS, THE
WHAT A PICTURE!: FACES AND FIGURES
WHAT A PICTURE!: IMAGINATION AND TECHNIQUE
WHAT A PICTURE!: LANDSCAPE AND LIGHT
WHAT A PICTURE!: LIGHTING AND COMPOSITION
WHAT A PICTURE!: MAKING A PICTURE
WHAT A PICTURE!: PERSPECTIVES OF SPACE
WHAT A PICTURE!: THE VITAL MOMENT
WHAT A PICTURE!: TRAVELLING LIGHT
WILDLIFE PHOTOGRAPHER AT WORK, A
WORLD OF PHOTOGRAPHY, THE
WORLD'S GREATEST PHOTOGRAPHY COURSE

Photography — Travel photography
BILL CHRISTENBERRY'S TRIP TO ALABAMA
CAMERA: EARLY PHOTOGRAPHY: SOLDIERS OF THE QUEEN
CAMERA: EARLY PHOTOGRAPHY: WORLD ON A PLATE
CARTES POSTALES VIDEO
J.A. MARTIN, PHOTOGRAPHER
PHOTOGRAPHER OF THE YEAR
PHOTOGRAPHERS OF THE AMERICAN WEST: 1860-1880
PIONEERS OF PHOTOGRAPHY: FAMOUS MEN AND FAIR WOMEN
PIONEERS OF PHOTOGRAPHY: TRAVELLING MAN
TERRITOIRES PHOTOGRAPHIQUES
VALLEE DE LA MORT, LA

Photography — Underwater photography
FROM THE LOG OF JUST LOVE
LAURIE SIMMONS: A TEASER
UNDER THE SEA BY PHOTOSPHERE

Photography — Vocational guidance
BUSINESS OF PHOTOGRAPHY, THE
PHOTOGRAPHIC VISION, THE: A LIFETIME IN PHOTOGRAPHY
PHOTOGRAPHIC VISION, THE: THE MARKETPLACE
PHOTOGRAPHY
PROFESSIONAL PHOTOGRAPHY
WAY I SEE IT, THE

Photography and dance
QUAI BOURBON

Photography and history
CAMERA: EARLY PHOTOGRAPHY: CAUGHT IN TIME
IMAGES OF GLORY
MATHEW BRADY: PHOTOGRAPHER OF AN ERA
PHOTOGRAPHERS OF THE AMERICAN WEST: 1860-1880
REMEMBERING LIFE
ROMANTIC HORIZON, THE
TRUE STORY OF THE CIVIL WAR

Photography and homosexuality
ROBERT MAPPLETHORPE

Photography and labor
AMERICA AND LEWIS HINE
AMERICAN IMAGES: PHOTOGRAPHS AND PHOTOGRAPHERS OF THE FARM SECURITY ADMINISTRATION, 1932-1942
DOCUMENTARILY THINE: LEWIS HINE
ENA'S ADVENTURES: PART ONE
HOMAGE TO AUGUST SANDER
PHOTOGRAPHER: RUSSELL LEE
RETURN JOURNEY
VOYAGE DE ROSE, UN

Photography and literature
A. DUMAS OU L'HYPNOSE CINEMA
ANDERSEN CHEZ LE PHOTOGRAPHE
AU PAYS DES VISAGES
CLIMATE OF NEW YORK, THE
HOLLIS FRAMPTON
MARINE TERRACE
STRINDBERG: PEINTRE ET PHOTOGRAPHE
ZOLA THE PHOTOGRAPHER

Photography and music
GUY LE QUERREC, PHOTOGRAPHE
PATRICK ULLMANN, PHOTOGRAPHE DES CHANTEURS FRANCAIS
WILLY RONIS OU LES CADEAUX DU HASARD

Photography and painting
CAMERA: EARLY PHOTOGRAPHY: ENDS OF ART

DAVID HOCKNEY'S DIARIES
DEGAS
DEGAS DELACROIX DU REEL A L'IMAGINAIRE
PORTRAITS
SIX FACES OF ROYALTY: VICTORIA
STRINDBERG: PEINTRE ET PHOTOGRAPHE
UNQUIET SPIRIT, THE: THE LIFE AND ART OF EDGAR DEGAS (1834-1917)
VICTORIAN PHOTOGRAPHY

Photography and politics
ANNEES KAGAN, LES
BILL CHRISTENBERRY'S TRIP TO ALABAMA
DAVID GOLDBLATT IN BLACK AND WHITE
DOCUMENTARY URGE, THE: TOM ARNDT
MARTIN CHAMBI AND THE HEIRS OF THE INCAS
PHOTOMONTAGE TODAY: PETER KENNARD
SENZENINA: WHAT HAVE WE DONE TO DESERVE THIS?

Photography and psychology
BILL PARKER

Photography and science
BEYOND THE OCEAN: BENEATH A LEAF
COULEUR ET PHOTO
EARLY PHOTOGRAPHERS, THE
KEEPERS OF THE LIGHT
MAREY: UN SAVANT TROP MECONNU
MOVING STILL
PHOTOGRAPHIE DEVANT LES BESOINS SPECIFIQUES DES ARCHEOLOGUES, LA
PIONEERS OF PHOTOGRAPHY: THE FLEETING IMAGE

Photography and society
AMERICA AND LEWIS HINE
AMERICAN IMAGES: PHOTOGRAPHS AND PHOTOGRAPHERS OF THE FARM SECURITY ADMINISTRATION, 1932-1942
ART ET LA DECADENCE, L'
BILL CHRISTENBERRY'S TRIP TO ALABAMA
BLACK HAS ALWAYS BEEN BEAUTIFUL
CAMERA: EARLY PHOTOGRAPHY: OUTBACK AND BEYOND
CAMERA: EARLY PHOTOGRAPHY: THE OTHER HALF
CONVERSATIONS WITH ROY DECARAVA
DOCUMENTARILY THINE: LEWIS HINE
DOCUMENTARY URGE, THE: TOM ARNDT
DOROTHEA LANGE
DOROTHEA LANGE: THE CLOSER FOR ME
DOROTHEA LANGE: UNDER THE TREES
EDWARD STEICHEN
ENFANCE D'UN PHOTOGRAPHE: LARTIGUE
GRAND REPORTER
IN THE STREET
JACQUES-HENRI LARTIGUE
JOHN ALBOK'S NEW YORK
"LET US NOW PRAISE FAMOUS MEN":
ALABAMA 40 YEARS ON
MARGARET MONCK
MARGINA 88
MILTON ROGOVIN
MILTON ROGOVIN
MOMENT IN TIME, A
MOVING STILLS
PHOTOGRAPHERS OF EVERYDAY LIFE
PHOTOGRAPHES ET CRINOLINES
PHOTOGRAPHIC VISION, THE: READING PHOTOGRAPHS
PHOTOGRAPHIE ET SOCIETE
PHOTOGRAPHY
POL MARA
QUELLE BELLE EPOQUE
RECITS D'ELLIS ISLAND: TRACES
REFLECTIONS
REPORTERS
ROBERT DOISNEAU, BADAUD DE PARIS, PECHEUR D'IMAGES
TIME MACHINE, THE
UNCOMMON IMAGES: THE HARLEM OF JAMES VAN DER ZEE
URSULA POWYS-LYBBE
VICTORIAN PHOTOGRAPHY
W. EUGENE SMITH: PHOTOGRAPHY MADE DIFFICULT
WEAPONS OF GORDON PARKS, THE
WORLD THAT DISAPPEARED, THE: THE PICTURES OF ROMAN VISHNIAC

Photography and technology
CONNECTIONS: 9-COUNTDOWN

Photography and war
AUSCHWITZ L'ALBUM LA MEMOIRE
CAMERA: EARLY PHOTOGRAPHY: EYEWITNESS OF WAR
CAMERA: EARLY PHOTOGRAPHY: SOLDIERS OF THE QUEEN
CIVIL WAR PHOTOGRAPHERS: WITNESS TO THE STORM
HANS STAUB, PHOTOJOURNALIST
HIROSHIMA BILDER
JOHN HOAGLAND: FRONTLINE PHOTOGRAPHER
LIFE AND DEATH OF PICTURE POST, THE
LIVES OF LEE MILLER, THE
MATHEW BRADY: PHOTOGRAPHER OF AN ERA
MOMENT IN TIME, A
PHOTOGRAPHER: RUSSELL LEE
PICTURING DERRY
SENZENINA: WHAT HAVE WE DONE TO DESERVE THIS?
STEICHEN: A CENTURY IN PHOTOGRAPHY
TOMORROW IS YOURS: DON MCCULLIN
TRUE STORY OF THE CIVIL WAR
VMCJ

Physically handicapped artists
ALBEDO
OLIVER GAGLIANI

Post-1945 art
ATELIER DE CHRISTIAN BOLTANSKI, L': 1983
BERNARD FAUCON, PHOTOGRAPHE
BERNARD FAUCON, PHOTOGRAPHE
BOLTANSKI
BOLTANSKI
CONTINUOUS DIARY, THE
GEORGES ROUSSE
GEORGES ROUSSE
JOHN BALDESSARI
JOHN BALDESSARI: A CONVERSATION
PEINTURES FRAICHES, MADE IN FRANCE
RICHARD PRINCE: EDITIONS
SOUVENIRS DE JEUNESSE
SUSAN HILLER: A PROVISIONAL PORTRAIT

THEATRE OF INDIFFERENCE
Schools — Art academies
NATHAN LYONS
Sculpture — Impressionist
UNQUIET SPIRIT, THE: THE LIFE AND ART OF EDGAR DEGAS (1834-1917)
Sculpture — Modernist
CORREO CULTURAL
FILM SUR HANS BELLMER
HANS BELLMER
L.E.A.R.: THE LEAGUE OF REVOLUTIONARY WRITERS AND ARTISTS
MOSCOU PARIS (1900-1930)
PISTA: THE MANY FACES OF STEPHEN DEUTCH
PORTRAITS
Sculpture — Post-1945
ARNULF RAINER
ARNULF RAINER
ARNULF RAINER
BILL CHRISTENBERRY'S TRIP TO ALABAMA
KENNETH SNELSON
PEINTURES FRAICHES, MADE IN FRANCE
SNOWSCREEN
Trompe l'oeil
GEORGES ROUSSE
Women artists
ART OF MARIE COSINDAS, THE
AU PAYS DES VISAGES
BARBARA MORGAN
BARBARA MORGAN: EVERYTHING IS DANCING
BETTY HAHN
CAMERA: EARLY PHOTOGRAPHY: MIRROR IMAGE
CARLOTTA CORPRON: DESIGNER WITH LIGHT
CINDY SHERMAN: AN INTERVIEW
CIRCLE OF LIGHT: THE PHOTOGRAPHY OF PAMELA BONE
CONTEMPORARY PHOTOGRAPHERS: RUTH BERNHARD
CONVERSATION WITH LOTTE JACOBI, A
DOROTHEA LANGE
DOROTHEA LANGE: THE CLOSER FOR ME
DOROTHEA LANGE: UNDER THE TREES
DREI LEBEN DER ILSE BING, DIE
EVON STREETMAN
FACE TO FACE
FAY GODWIN
FRIDA KAHLO AND TINA MODOTTI
GISELE FREUND: 1982
GOING WHERE I'VE NEVER BEEN: PHOTOGRAPHY OF DIANE ARBUS
GRACE ROBERTSON
HELEN MUSPRATT
IMOGEN CUNNINGHAM AT 93
IMOGEN CUNNINGHAM, PHOTOGRAPHER
L.E.A.R.: THE LEAGUE OF REVOLUTIONARY WRITERS AND ARTISTS
LAURA GILPIN: AN ENDURING GRACE
LAURIE SIMMONS: A TEASER
LENSES ON NATURE
LIVES OF LEE MILLER, THE
MARCIA RESNICK
MARGARET BOURKE-WHITE
MARGARET MONCK
NEVER GIVE UP — IMOGEN CUNNINGHAM
OTHER OBSERVERS, THE: WOMEN PHOTOGRAPHERS FROM 1850 TO THE PRESENT DAY
PHOTOGRAPHERS, THE
PHOTOGRAPHIC VISION, THE: COLOR
PHOTOGRAPHIE ET SOCIETE
PHOTOGRAPHY EXHIBITIONS
PHOTOPROFILES
PIONEERS OF PHOTOGRAPHY: FAMOUS MEN AND FAIR WOMEN
PORTRAIT OF IMOGEN
RUTH BERNHARD
SOUTH-EAST NUBA, THE
SUSAN HILLER: A PROVISIONAL PORTRAIT
SVET ZVANY FESTIVAL
TWO PHOTOGRAPHERS: WYNN BULLOCK, IMOGEN CUNNINGHAM
ULYSSE
URSULA POWYS-LYBBE
WAY I SEE IT, THE
WOMAN BEHIND THE IMAGE, THE: PHOTOGRAPHER JUDY DATER

NAMES INDEX

Abramowitz, Jerry
PHOTOGRAPHIC VISION, THE: THE STUDIO
Adams, Ansel Easton
ANSEL ADAMS
ANSEL ADAMS: PHOTOGRAPHER
ANSEL ADAMS: PHOTOGRAPHER
ANSEL ADAMS: PHOTOGRAPHER OF THE AMERICAN WEST
LANGUAGE OF THE CAMERA EYE
PHOTOGRAPHERS OF EVERYDAY LIFE
PHOTOGRAPHIC VISION, THE: LANDSCAPE AND CITYSCAPE
PHOTOGRAPHY AS AN ART
POINTS OF VIEW
PROFESSIONAL PHOTOGRAPHY
TECHNIQUE
Adams, Eddie
DAVID HARTMAN: THE SHOOTERS
Adamson, Robert
IN CAMERA
PIONEERS OF PHOTOGRAPHY: SUN PICTURES
Agathe Gaillard gallery (Paris, France)
PARIS 1985: LA GALERIE DE PHOTOS D'AGATHE GAILLARD
Aigner, Lucien
LUCIEN AIGNER: A LIFE WITH A CAMERA
Albok, John
JOHN ALBOK'S NEW YORK
Alvarez Bravo, Manuel
MANUEL ALVAREZ BRAVO, FOTOGRAFO
Andersen, Hans Christian
ANDERSEN CHEZ LE PHOTOGRAPHE
Arbus, Diane
GOING WHERE I'VE NEVER BEEN: PHOTOGRAPHY OF DIANE ARBUS
Arles — Rencontres Internationales de la Photographie (Arles, France)
J'AIME A LA FOLIE...LA PHOTOGRAPHIE
JEAN DIEUZAIDE: 1982
RENCONTRES PHOTOGRAPHIQUES D'ARLES, LES: 1981
Arndt, Thomas
DOCUMENTARY URGE, THE: TOM ARNDT
Atget, Jean Eugène Auguste
ATGET
EUGENE ATGET
EUGENE ATGET: PHOTOGRAPHER
Baer, Morley
CONTEMPORARY PHOTOGRAPHERS: MORLEY BAER
MORLEY BAER
Bailey, David Royston
DAVID BAILEY
Baldessari, John Anthony
JOHN BALDESSARI
JOHN BALDESSARI: A CONVERSATION
Bannister, Edward Mitchell
SHARING TRADITIONS: FIVE BLACK ARTISTS IN NINETEENTH-CENTURY AMERICA
Barbieri, Gianpaolo
GIANPAOLO BARBIERI
Barceló, Miguel
PORTRAIT D'UN PORTRAITISTE
Baudelaire, Charles
DYNAMO: SPECIAL PHOTO
Bayard, Hippolyte
PIONEERS OF PHOTOGRAPHY: SUN PICTURES
Beach, Claude
LOCAL STRANGER, A
Beksinski, Zdzislaw
PHOTORAMA
Bellmer, Hans
FILM SUR HANS BELLMER
HANS BELLMER
Beloff, Halla
TIME AND LIGHT
Bentley, William
CRYSTALS
Berger, John
TIME AND LIGHT
Bernhard, Ruth
CONTEMPORARY PHOTOGRAPHERS: RUTH BERNHARD
RUTH BERNHARD
Berry, Ian
SOCIAL DOCUMENTARY
Bing, Ilse
DREI LEBEN DER ILSE BING, DIE
Blakemore, John
LANDSCAPE
Boltanski, Christian
ATELIER DE CHRISTIAN BOLTANSKI, L': 1983
BOLTANSKI
BOLTANSKI
DYNAMO: SPECIAL PHOTO
SOUVENIRS DE JEUNESSE

Bone, Pamela
CIRCLE OF LIGHT: THE PHOTOGRAPHY OF PAMELA BONE
Bourke-White, Margaret
MARGARET BOURKE-WHITE
Bourne, Samuel
CAMERA: EARLY PHOTOGRAPHY: WORLD ON A PLATE
PIONEERS OF PHOTOGRAPHY: TRAVELLING MAN
Brady, Mathew B.
MATHEW BRADY: PHOTOGRAPHER OF AN ERA
TRUE STORY OF THE CIVIL WAR
Brandt, Bill
BILL BRANDT
SHADOWS FROM LIGHT
Brassaï (aka Gyula Halász)
BRASSAI
Breakwell, Ian
CONTINUOUS DIARY, THE
Breitner, George Hendrik
GEORGE HENDRIK BREITNER
Brihat, Denis
RENCONTRE AVEC JEAN-PIERRE SUDRE ET DENIS BRIHAT
Bullock, Wynn
TWO PHOTOGRAPHERS: WYNN BULLOCK, IMOGEN CUNNINGHAM
WYNN BULLOCK: PHOTOGRAPHER
Burckhardt, Rudy
CLIMATE OF NEW YORK, THE
RUDY BURCKHARDT
Burrows, Larry
PHOTOGRAPHERS, THE
Callahan, Harry Morey
HARRY CALLAHAN: A BROADCAST EXHIBITION
HARRY CALLAHAN: A NEED TO SEE AND EXPRESS
HARRY CALLAHAN: ELEANOR AND BARBARA
Cameron, Julia Margaret
CAMERA: EARLY PHOTOGRAPHY: MIRROR IMAGE
PHOTOGRAPHIC VISION, THE: A VISUAL HERITAGE
PIONEERS OF PHOTOGRAPHY: FAMOUS MEN AND FAIR WOMEN
Capa, Cornell
CORNELL CAPA, PARTS I & II
Carroll, Lewis (aka Charles Lutwidge Dodgson)
CAMERA: EARLY PHOTOGRAPHY: MIRROR IMAGE
Cartier-Bresson, Henri
CONTACTS
EYE OF CARTIER-BRESSON, THE
FLAGRANTS DELITS
PHOTOGRAPHERS OF EVERYDAY LIFE
QUEBEC VU PAR CARTIER-BRESSON, LE
Cavagnard, David
CONTEMPORARY PHOTOGRAPHERS: DAVID CAVAGNARD
Chaimowicz, Marc Camille
TIME AND LIGHT
Chambi, Martin
MARTIN CHAMBI AND THE HEIRS OF THE INCAS
Chi, Tseng Kwong
EAST MEETS WEST
Christenberry, William
BILL CHRISTENBERRY'S TRIP TO ALABAMA
Ciezlewicz, Roman
ROMAN: PHOTO MONTEUR
Clergue, Lucien
PHOTOGRAPHIE AMATEUR, LA
PHOTOGRAPHIE, LA: TECHNIQUES
Coleman, Allan Douglas
A.D. COLEMAN
Connor, Linda
PHOTOGRAPHIC VISION, THE: CONTROLLING EXPOSURE
PHOTOGRAPHIC VISION, THE: LANDSCAPE AND CITYSCAPE
Converse, Gordon
ONE SEEING...
Corpron, Carlotta M.
CARLOTTA CORPRON: DESIGNER WITH LIGHT
Cosindas, Marie
ART OF MARIE COSINDAS, THE
Cowin, Eileen
PHOTOGRAPHIC VISION, THE: SEEING WITH THE CAMERA
Criswell, Ronnie
OURSELVES AND THAT PROMISE
Crouch, Steve
CONTEMPORARY PHOTOGRAPHERS: STEVE CROUCH
Cunningham, Imogen
IMOGEN CUNNINGHAM AT 93
IMOGEN CUNNINGHAM, PHOTOGRAPHER
NEVER GIVE UP—IMOGEN CUNNINGHAM
PORTRAIT OF IMOGEN
TWO PHOTOGRAPHERS: WYNN BULLOCK, IMOGEN CUNNINGHAM
Curtis, Edward Sheriff
SHADOW CATCHER, THE
WALKING IN A SACRED MANNER
Daguerre, Louis Jacques Mandé
CAMERA: EARLY PHOTOGRAPHY: FIRST IMPRESSIONS
DAGUERRE OU LA NAISSANCE DE LA PHOTOGRAPHIE
INVENTION DE LA PHOTOGRAPHIE, L'
KEEPERS OF THE LIGHT
PIONEERS OF PHOTOGRAPHY: MIRROR WITH A MEMORY
Dater, Judy
PHOTOGRAPHIC VISION, THE: THE PORTRAIT
WOMAN BEHIND THE IMAGE, THE: PHOTOGRAPHER JUDY DATER
Davis, Billy
OURSELVES AND THAT PROMISE
DeCarava, Roy
CONVERSATIONS WITH ROY DECARAVA
Degas, Hilaire Germain Edgar
DEGAS
DEGAS DELACROIX DU REEL A L'IMAGINAIRE
UNQUIET SPIRIT, THE: THE LIFE AND ART OF EDGAR DEGAS (1834-1917)
Delacroix, Eugène
DEGAS DELACROIX DU REEL A L'IMAGINAIRE
Deluc, Michel
PHOTOSCOPIES
Depardon, Raymond
ANNEES-DECLIC, LES
CONTACTS
REPORTERS
Deutch, Stephen
PISTA: THE MANY FACES OF STEPHEN DEUTCH
Dieuzaide, Jean
JEAN DIEUZAIDE: 1982
PORTRAIT DE JEAN DIEUZAIDE

Divola, John
PHOTOGRAPHIC VISION, THE: CONTROLLING EXPOSURE
Doisneau, Robert
CONTACTS
ECRIVAINS PHOTOGRAPHIES PAR ROBERT DOISNEAU, LES
ENTREE LIBRE
OBJECTIF VILLE: ROBERT DOISNEAU
PARIS OF ROBERT DOISNEAU, THE
ROBERT DOISNEAU, BADAUD DE PARIS, PECHEUR D'IMAGES
Dominis, John
PHOTOGRAPHERS, THE
Du Camp, Maxime
PIONNIERS, LES
Ducrot, Jérôme
PHOTOSCOPIES
Duffy
PORTRAIT, THE
Dumas, Alexandre
A. DUMAS OU L'HYPNOSE CINEMA
Duncanson, Robert Scott
SHARING TRADITIONS: FIVE BLACK ARTISTS IN NINETEENTH-CENTURY AMERICA
Durgen, Ken
WILDLIFE PHOTOGRAPHER AT WORK, A
Eames, Charles
SOMETHING ABOUT PHOTOGRAPHY
Eastman, George
GEORGE EASTMAN: ONE MAN'S VISION, IMAGE FOR ALL
Eisenstaedt, Alfred
ALFRED EISENSTAEDT
EISENSTAEDT—GERMANY
PHOTOGRAPHERS, THE
Elisofon, Eliot
TRIBUTE TO AFRICA
Emerson, Peter Henry
CAMERA: EARLY PHOTOGRAPHY: TRUTH AND BEAUTY
Erwitt, Elliott
CONTACTS
Evans, Walker
"LET US NOW PRAISE FAMOUS MEN": ALABAMA 40 YEARS ON
WALKER EVANS: HIS TIME—HIS PRESENCE—HIS SILENCE
Evergon (aka Celluloso Evergonni; Eve R. Gonzalez; Egon Brut)
JUST EVERGON
Faucon, Bernard
BERNARD FAUCON
BERNARD FAUCON, PHOTOGRAPHE
BERNARD FAUCON, PHOTOGRAPHE
Feininger, Andreas
ANDREAS FEININGER
ANDREAS FEININGER
Fellman, Sandi
PHOTOGRAPHIC VISION, THE: THE STUDIO
Fenton, Roger
CAMERA: EARLY PHOTOGRAPHY: EYEWITNESS OF WAR
Finn, David
THEATRE OF INDIFFERENCE
Fischer, Hal
HAL FISCHER READS *AMERICAN PHOTOGRAPHER:* T & A MADE EASY
Flick, Robbert
PHOTOGRAPHIC VISION, THE: VISUAL DOCUMENTS
Fokkema, Gerrit
FACE TO FACE
Fox Talbot Museum (Wiltshire, England)
PLACE IN THE COUNTRY, A: LACOCK ABBEY, WILTSHIRE
Frampton, Hollis
HOLLIS FRAMPTON
Frank, Robert Louis
FIRE IN THE EAST: A PORTRAIT OF ROBERT FRANK
ROBERT FRANK: L'IMMAGINE DELLA POESIA
Fresson, Michel
PROCEDE FRESSON, LE
Freund, Gisèle
AU PAYS DES VISAGES
GISELE FREUND: 1982
PHOTOGRAPHIE ET SOCIETE
Friedman, Benno
BENNO FRIEDMAN
Gagliani, Oliver L.
OLIVER GAGLIANI
Gamma Agency (Paris, France)
REPORTERS
Gassmann, Pierre
RUE DU REGARD, UN PORTRAIT DE PIERRE GASSMANN
Gibson, Dan
SOLITUDES
Gidal, Tim
MASS AGE IMAGE, THE: PHOTOJOURNALISM; PART I: THE BEGINNINGS OF PHOTOJOURNALISM: ERICH SALOMON; PART II: THE ESSENCE: TIM GIDAL
Gilpin, Laura
LAURA GILPIN: AN ENDURING GRACE
Godwin, Fay
FAY GODWIN
Goldblatt, David
DAVID GOLDBLATT IN BLACK AND WHITE
Gowin, Emmet
PHOTOGRAPHER'S EYE, THE
Gremp, David
DAVID GREMP: JESSICA, JULY 15, 1983
Griffin, Brian
PORTRAIT, THE
Haas, Ernst
LANDSCAPE
MOVEMENT
Hahn, Betty
BETTY HAHN
Halberstadt, Milton
PROFESSIONAL PHOTOGRAPHY
Hamilton, David
DAVID HAMILTON
Hardy, Bert
BERT HARDY
LIFE AND DEATH OF PICTURE POST, THE
Harmon, Byron
GREAT DAYS IN THE ROCKIES
Harris, Brian
PHOTOJOURNALISM
Hartung, Hans
DU GESTE A L'INFINI: HANS HARTUNG
HARTUNG
Haskins, Sam
SAM HASKINS
SAM HASKINS
Heartfield, John
JOHN HEARTFIELD: FILM UEBER DEN FOTOMONTEUR
Heinecken, Robert
PHOTOGRAPHY AS FINE ART
ROBERT HEINECKEN
Hill, David Octavius
IN CAMERA
PIONEERS OF PHOTOGRAPHY: SUN PICTURES
Hill, Oliver Massey
FIXED IN TIME: A VICTORIAN ALBUM

Hill, Paul
IMAGE, THE
Hiller, Susan
SUSAN HILLER: A PROVISIONAL PORTRAIT
Hine, Lewis Wickes
AMERICA AND LEWIS HINE
DOCUMENTARILY THINE: LEWIS HINE
RECITS D'ELLIS ISLAND: TRACES
Hoagland, John
JOHN HOAGLAND: FRONTLINE PHOTOGRAPHER
Hockney, David
DAVID HOCKNEY'S DIARIES
HOCKNEY THE PHOTOGRAPHER
Horst, Horst P.
HORST
Howard, David
DAVID HOWARD: AN ARTIST'S MEDIUM AND METHODS
Hugo, Victor Marie
MARINE TERRACE
Jacobi, Lotte
CONVERSATION WITH LOTTE JACOBI, A
Jammes, Louis
LOUIS JAMMES
Joffet, Emmanuel
TOUR DE L'AFRIQUE EN QUATRE-VINGTS JOURS, LE
Johnston, Joshua
SHARING TRADITIONS: FIVE BLACK ARTISTS IN NINETEENTH-CENTURY AMERICA
Johnstone, Mark
PHOTOGRAPHIC VISION, THE: THE CAMERA
Josephson, Kenneth
KENNETH JOSEPHSON
Jussim, Estelle
ESTELLE JUSSIM
Kagan, Elie
ANNEES KAGAN, LES
Kahlo, Frida
FRIDA KAHLO AND TINA MODOTTI
Karsh, Yousuf
KARSH: THE SEARCHING EYE
Kasten, Barbara
PHOTOGRAPHIC VISION, THE: STRUCTURE WITHIN THE IMAGE
PHOTOGRAPHIC VISION, THE: THE CAMERA
Kennard, Peter
PHOTOMONTAGE TODAY: PETER KENNARD
Kennerly, David
DAVID HARTMAN: THE SHOOTERS
Kertész, André
ANDRE DANS LES VILLES: BUDAPEST, PARIS, NEW YORK, 1894-1985: PORTRAIT D'ANDRE KERTESZ
ANDRE KERTESZ
ANDRE KERTESZ, PHOTOGRAPHE AMERICAIN A PARIS: 1981
ANDRE KERTESZ: A POET WITH A CAMERA
EVERYTHING IS PHOTOGRAPH: PROFILE OF ANDRE KERTESZ
MUSIC IS THE STORY
Killip, Chris
URBAN LANDSCAPE
Kirkland, Douglas
DAVID HARTMAN: THE SHOOTERS
Klein, William
CONTACTS
WILLIAM KLEIN: PAINTER, PHOTOGRAPHER, FILMMAKER
Kodak Museum (Bradford, England)
KODAK MUSEUM
Koudelka, Josef
CONTACTS
Labrot, Syl W.
SYL LABROT
Land, Edwin Herbert
COLOURFUL NOTIONS
Lange, Dorothea
DOROTHEA LANGE
DOROTHEA LANGE: THE CLOSER FOR ME
DOROTHEA LANGE: UNDER THE TREES
Lartigue, Jacques-Henri
ENFANCE D'UN PHOTOGRAPHE: LARTIGUE
EXTRAITS DU JOURNAL DE JACQUES-HENRI LARTIGUE
JACQUES-HENRI LARTIGUE
JACQUES-HENRI LARTIGUE
JACQUES-HENRI LARTIGUE, PEINTRE ET PHOTOGRAPHE
JACQUES-HENRI LARTIGUE: 1980
Le Diascorn, François
PHOTOSCOPIES
Le Querrec, Guy
GUY LE QUERREC, PHOTOGRAPHE
PHOTOSCOPIES
Lee, Russell
PHOTOGRAPHER: RUSSELL LEE
Leifer, Neil
DAVID HARTMAN: THE SHOOTERS
Lewis, Edmonia
SHARING TRADITIONS: FIVE BLACK ARTISTS IN NINETEENTH-CENTURY AMERICA
Lichfield, Patrick
LICHFIELD ON PHOTOGRAPHY
Life magazine (New York, New York)
REMEMBERING LIFE
Lindbergh, Peter
PETER LINDBERGH
Livernois, Jules-Ernest
ERNEST LIVERNOIS, PHOTOGRAPHE
Lumière, Antoine
LUMIERE PERE ET FILS
Lumière, Auguste
LUMIERE PERE ET FILS
Lumière, Louis
LUMIERE PERE ET FILS
Lutz, Paul
PAUL LUTZ
Lyons, Nathan
NATHAN LYONS
PHOTOGRAPHIC VISION, THE: ARTISTIC EXPRESSION
PHOTOGRAPHIC VISION, THE: READING PHOTOGRAPHS
PHOTOGRAPHIC VISION, THE: VISUAL DOCUMENTS
Mandel, Mike
MANDEL/SULTAN
PHOTOGRAPHIC VISION, THE: READING PHOTOGRAPHS
Mapplethorpe, Robert
ROBERT MAPPLETHORPE
Mara, Pol
POL MARA
Marey, Jules Etienne
CAMERA: EARLY PHOTOGRAPHY: WHEEL OF LIFE—MOVEMENT
CE QUI ME MEUT
INTERMITTENCES: E.J. MAREY 1830-1904
MAREY: UN SAVANT TROP MECONNU
PIONEERS OF PHOTOGRAPHY: THE FLEETING IMAGE
Mark, Mary Ellen
DAVID HARTMAN: THE SHOOTERS
PHOTOGRAPHERS, THE

Marvin, David
ALBEDO
McCullin, Donald
SOCIAL DOCUMENTARY
TOMORROW IS YOURS: DON MCCULLIN
Mehta, Dilip
TRAVELLING LIGHT: THE PHOTOJOURNALISM OF DILIP MEHTA
Meyerowitz, Joel
PORTRAITS
URBAN LANDSCAPE
Michals, Duane
DUANE MICHALS (1939-1997)
Michel, Pierre
PIERRE MICHEL!!! ORCHESTRE A FICTIONS!!!
Miller, Lee
LIVES OF LEE MILLER, THE
Misonne, Léonard
LEONARD MISONNE: THE PHOTOGRAPHER
Modotti, Tina
FRIDA KAHLO AND TINA MODOTTI
Monck, Margaret
MARGARET MONCK
Morgan, Barbara
BARBARA MORGAN
BARBARA MORGAN: EVERYTHING IS DANCING
Mudford, Grant
PHOTOGRAPHIC VISION, THE: SEEING WITH THE CAMERA
Musée Français de la Photographie (Bièvres, France)
FRENCH MUSEUM OF PHOTOGRAPHY, THE
Muspratt, Helen
HELEN MUSPRATT
Muybridge, Eadweard
ANIMALS IN MOTION
CAMERA: EARLY PHOTOGRAPHY: WHEEL OF LIFE—MOVEMENT
CHEVAL DE FER, LE
DEGAS
EADWEARD MUYBRIDGE, ZOOPRAXOGRAPHER
FILMING MUYBRIDGE
MUYBRIDGE IN MOTION
MUYBRIDGE REVISITED
PIONEERS OF PHOTOGRAPHY: THE FLEETING IMAGE
PIONNIERS, LES
Nadar (aka Gaspard Félix Tournachon)
DYNAMO: SPECIAL PHOTO
FRENCH MUSEUM OF PHOTOGRAPHY, THE
GLOIRE A FELIX TOURNACHON
NADAR
PHOTOGRAPHIC VISION, THE: A VISUAL HERITAGE
PIONEERS OF PHOTOGRAPHY: NADAR THE GREAT
PIONNIERS, LES
VIE D'AVENTURES, UNE: FELIX NADAR
Namuth, Hans
HANS NAMUTH, INSTANTS DONNES
PHOTOGRAPHY EXHIBITIONS
National Museum of Photography, Film and Television (Bradford, England)
KODAK MUSEUM
Nègre, Charles
QUAI BOURBON
Newhall, Beaumont
LANGUAGE OF THE CAMERA EYE
Newman, Arnold
PHOTOGRAPHIC VISION, THE: THE PORTRAIT
PORTRAIT, THE
Newton, Helmut
CONTACTS
HELMUT NEWTON
Niépce, Joseph Nicéphore
HERITAGE FABULEUX DE NICEPHORE NIEPCE, L'
INVENTION DE LA PHOTOGRAPHIE, L'
KEEPERS OF THE LIGHT
PIONEERS OF PHOTOGRAPHY: MIRROR WITH A MEMORY
North, Kenda
PHOTOGRAPHIC VISION, THE: COLOR
Parker, Bill
BILL PARKER
Parkinson, Norman
FASHION IN THE MAKING
PHOTOGRAPHERS, THE
Parks, Gordon Alexander Buchanan
GORDON PARKS: MOMENTS WITHOUT PROPER NAMES
WEAPONS OF GORDON PARKS, THE
Patterson, Freeman
FREEMAN IN TIME
Pearlstein, Philip
PORTRAITS
Picture Post magazine (London, England)
LIFE AND DEATH OF PICTURE POST, THE
Pollock, George
IMAGE OF LIGHT: THE PHOTOGRAPHY OF GEORGE POLLOCK
Porter, Eliot Furness
ELIOT PORTER
ELIOT PORTER'S WORLD
Powys-Lybbe, Ursula
URSULA POWYS-LYBBE
Prince, Richard
RICHARD PRINCE: EDITIONS
Rainer, Arnulf
ARNULF RAINER
ARNULF RAINER
ARNULF RAINER
HIROSHIMA BILDER
Rankaitis, Susan
PHOTOGRAPHIC VISION, THE: ARTISTIC EXPRESSION
Rapho agency (France)
RAPHO, HISTOIRE D'UNE FAMILLE
Rauschenberg, Robert
PHOTOGRAPHY EXHIBITIONS
Ray, Man
BANDE A MAN RAY, LA
MAN RAY
MAN RAY
MAN RAY
MAN RAY PHOTOGRAPHE
VISITE DE L'ATELIER DE MAN RAY, UNE
Resnick, Marcia Aylene
MARCIA RESNICK
Riefenstahl, Leni
SOUTH-EAST NUBA, THE
Riger, Robert
ATHLETE, THE: ROBERT RIGER'S DRAWINGS, PHOTOGRAPHS AND TELEVISION SPORTS
Robertson, Grace
GRACE ROBERTSON
Rogovin, Milton
MILTON ROGOVIN
MILTON ROGOVIN
Ronis, Willy
JOURNEE AVEC WILLY RONIS, UNE
PHOTOGRAPHE WILLY RONIS, LE
VOYAGE DE ROSE, UN
WILLY RONIS OU LES CADEAUX DU HASARD

Rousse, Georges
GEORGES ROUSSE
GEORGES ROUSSE
Russell, Andrew Joseph
ANDREW J. RUSSELL, A VISUAL HISTORIAN
Salgado, Sebastiao
CONTACTS
Salomon, Erich
MASS AGE IMAGE, THE: PHOTOJOURNALISM; PART I: THE BEGINNINGS OF PHOTOJOURNALISM: ERICH SALOMON; PART II: THE ESSENCE: TIM GIDAL
Sander, August
HOMMAGE A AUGUST SANDER
Sandved, Kjell
BEYOND THE OCEAN: BENEATH A LEAF
Scharf, David
PHOTOGRAPHIC VISION, THE: VISUAL DOCUMENTS
Sherman, Cindy
CINDY SHERMAN: AN INTERVIEW
PHOTOGRAPHY EXHIBITIONS
Sieff, Jeanloup
J.L. SIEFF
JEANLOUP SIEFF
PHOTOSCOPIES
VALLEE DE LA MORT, LA
Simmons, Laurie
LAURIE SIMMONS: A TEASER
Sisic, Sahin
MARGINA 88
Siskind, Aaron
AARON SISKIND
Slavin, Neal
NEAL SLAVIN
Smith, Derek
RETURN JOURNEY
Smith, Keith A.
KEITH SMITH
Smith, Michael A.
WORLD OF PHOTOGRAPHY, THE
Smith, W. Eugene
W. EUGENE SMITH: PHOTOGRAPHY MADE DIFFICULT
Snelson, Kenneth D.
KENNETH SNELSON
KENNETH SNELSON: THE PHOTOGRAPHS
Snow, Michael
SNOWSCREEN
Snowdon, Lord (aka Anthony Charles Robert Armstrong-Jones)
SNOWDON ON CAMERA: PARTS 1 AND 2
Spence, Jo
TIME AND LIGHT
Spender, John Humphrey
RETURN JOURNEY
Stackpole, Peter
BRIDGEMEN, THE
Stanfield, Jim
PHOTOGRAPHER OF THE YEAR
Staub, Hans
HANS STAUB, FOTOREPORTER
Steichen, Edward
EDWARD STEICHEN
EDWARD STEICHEN
FAMILY OF MAN, THE
MASTER OF THE CAMERA: EDWARD J. STEICHEN
PHOTOGRAPHER, THE
STEICHEN: A CENTURY IN PHOTOGRAPHY
THIS IS EDWARD STEICHEN
Steiner, Ralph
RALPH STEINER
Stieglitz, Alfred
ALFRED STIEGLITZ: PHOTOGRAPHER (1864-1946)
Still, James
OURSELVES AND THAT PROMISE
Streetman, Evon
EVON STREETMAN
Strindberg, Johan August
STRINDBERG: PEINTRE ET PHOTOGRAPHE
Sudre, Jean-Pierre
RENCONTRE AVEC JEAN-PIERRE SUDRE ET DENIS BRIHAT
Sultan, Larry
MANDEL/SULTAN
PHOTOGRAPHIC VISION, THE: READING PHOTOGRAPHS
Sutcliffe, Frank Meadow
CAMERA: EARLY PHOTOGRAPHY: TRUTH AND BEAUTY
Sygma Agency (France)
SYGMA: UNE AGENCE PHOTO: 1980
Talbot, William Henry Fox
CAMERA: EARLY PHOTOGRAPHY: FIRST IMPRESSIONS
KEEPERS OF THE LIGHT
LITTLE BIT OF MAGIC REALISED, A
PIONEERS OF PHOTOGRAPHY: THE PENCIL OF NATURE
PLACE IN THE COUNTRY, A: LACOCK ABBEY, WILTSHIRE
Tanner, Henry Ossawa
SHARING TRADITIONS: FIVE BLACK ARTISTS IN NINETEENTH-CENTURY AMERICA
Tessier, Albert
ALBERT TESSIER A FORCE D'IMAGES
Thomson, John
CAMERA: EARLY PHOTOGRAPHY: WORLD ON A PLATE
Tice, George A.
GEORGE TICE
Tingaud, Jean Marc
DERNIER ADIEU, DE JEAN MARC TINGAUD
Trager, Philip
PHILIP TRAGER
Turnley, David C.
SENZENINA: WHAT HAVE WE DONE TO DESERVE THIS?
Ullmann, Patrick
PATRICK ULLMANN, PHOTOGRAPHE DES CHANTEURS FRANCAIS
Van Der Zee, James Augustus Joseph
BLACK HAS ALWAYS BEEN BEAUTIFUL
UNCOMMON IMAGES: THE HARLEM OF JAMES VAN DER ZEE
Van Dyke, Willard Ames
CONVERSATIONS WITH WILLARD VAN DYKE
WILLARD VAN DYKE
Varda, Agnès
ULYSSE
Vishniac, Roman
WORLD THAT DISAPPEARED, THE: THE PICTURES OF ROMAN VISHNIAC
Voyeux, Martine
PHOTOSCOPIES
Wagner, Catherine
PHOTOGRAPHIC VISION, THE: LANDSCAPE AND CITYSCAPE
Waite, Charlie
LANDSCAPE PHOTOGRAPHER
Warren, Robert Penn
OURSELVES AND THAT PROMISE
Watkins, Carleton Emmons
CARLETON WATKINS
WET PLATE PHOTOGRAPHY PROCESS, THE

Weber, Al
CONTEMPORARY PHOTOGRAPHERS: AL WEBER
Weegee (aka Arthur Fellig)
REAL WEEGEE, THE
WEEGEE'S NEW YORK
Wegman, William
WILLIAM WEGMAN
WORLD OF PHOTOGRAPHY, THE
Welpott, Jack
PHOTOGRAPHIC VISION, THE: LANDSCAPE AND CITYSCAPE
PHOTOGRAPHIC VISION, THE: RESPONDING TO LIGHT
Weston, Edward
DAYBOOKS OF EDWARD WESTON, THE: HOW YOUNG I WAS
DAYBOOKS OF EDWARD WESTON, THE: THE STRONGEST WAY OF SEEING
EDWARD WESTON: ARTIST OF THE CAMERA
PHOTOGRAPHER, THE
Williams, John
FACE TO FACE
Williamson, John
UNDER THE SEA BY PHOTOSPHERE
Winogrand, Garry
PHOTOGRAPHER'S EYE, THE
PHOTOGRAPHIC VISION, THE: STRUCTURE WITHIN THE IMAGE
PHOTOGRAPHIC VISION, THE: THE CAMERA
Witkiewicz, Stanislaw Ignacy
PORTRAITS ET GRIMACES: WITKIEWICZ
Wood, Jim
PHOTOGRAPHIC VISION, THE: THE FILM
Yavno, Max
PHOTOGRAPHIC VISION, THE: A LIFETIME IN PHOTOGRAPHY
Zachmann, Patrick
DYNAMO: SPECIAL PHOTO
Zahalka, Anne
FACE TO FACE
Zola, Emile
ZOLA PHOTOGRAPHE
Zorn, Len
LENSES ON NATURE
Zorn, Maria
LENSES ON NATURE

SOURCE INDEX

Acmé Films
24-26 rue Paul Fort
Paris 75014
France
1.45.43.66.57

Acrobat Book Publishers
P.O.Box 480820
Los Angeles CA 90048
United States

Adler Video Marketing Ltd.
6849 Old Dominion Dr., Suite 360
McLean VA 22101
United States
703-556-8880
703-556-9288 (fax)

Agency for Instructional Technology
Box A
Bloomington IN 47402
United States
800-457-4509; 812-339-2203

Albany Video
The Albany, Douglas Way
London S.E.8
Great Britain
1.692.0231

Albright-Knox Art Gallery
1285 Elmwood Avenue
Buffalo NY 14222
United States
716-882-8700

Amarillo Art Center
P.O.Box 447 - 220 S Van Buren
Amarillo TX 79178
United States
806-371-5050; 806-372-8356

Ambassade du Danemark
77 avenue Marceau
Paris 75016
France
1.47.23.54.20

Ambrose Video Publishing
381 Park Ave. S. Suite 1601
New York NY 10016
United States
800-526-4663; 212-696-4545

American Federation of Arts
41 E. 65th St.
New York NY 10021
United States
212-988-7700
212-861-2487 (fax)

American Masters
15 West 26th St. 10 fl.
New York NY 10010
United States
212-532-2744

Animation Art Graphique Audiovisuel
71 faubourg Saint-Antoine
Paris 75011
France
1.43.43.57.06

Anomaly Films
135 Hudson St.
New York NY 10013
United States
212-925-1500
212-925-1501 (fax)

Antenne 2
22 avenue Montaigne
Paris Cedex 08 75008
France
1.42.99.42.42; 1.43.22.16.31; 1.44.21.42.42

Appalshop Films
306 Madison St.
Whitesburg KY 41858
United States
606-633-0108; 606-633-5708

Argos Films
4 rue Édouard-Nortier
Neuilly-sur-Seine 92200
France
1.47.22.91.26

Arts Council of Great Britain
105 Piccadilly
London W1V OAU
Great Britain
1.629.9495

Arts Council of Great Britain
Film Library
201 Felixstowe Rd.
Ipswich Suffolk 1P3 9BJ
Great Britain

artsAmerica
12 Havemeyer Place
Greenwich CT 06830
United States
203-869-4693; 203-869-4694

Association Française pour la Diffusion du Patrimoine Photographique
No address known
France

Association pour la Diffusion de la Pensée Française
9 rue Anatole de la Forge
Paris 75017
France
1.42.27.32.97

Australian Broadcasting Commission
145 Elizabeth St., P.O. Box 487 G.P.O
Sydney NSW 2001
Australia

Aylmer Press
P.O. Box 2735
Madison WI 53701
United States
603-233-2259

Baker & Taylor Video
501 S. Gladiolus
Momence IL 60954
United States
800-435-5111; 800-892-1892 (in IL)

BBC Enterprises
80 Wood Lane
London W12 OTT
Great Britain
1.743.5588; 1.576.0202; 212-581-7100

BBC Enterprises Library Sales
Windmill Road
Brentford
Middlesex TW8 9NF
Great Britain
1.568.7986; 212-581-7100

Beacon Films
930 Pitner Ave.
Evanston IL 60202
United States
800-323-5448; 312-328-6700

Bien Vu
12 rue Heinrich
Boulogne 92100
France
1.46.08.29.84

Biograph Entertainment, Ltd.
300 Phillips Park Rd. P.O.Box 190
Mamaroneck NY 10543
United States
914-381-2993

Blackwood Productions Inc.
251 West 57 Street, Suite 427
New York NY 10019
United States
212-247-4710

Brigham Young University, Audio Visual Services
211 MPS
Provo UT 84602
United States
801-378-1211

Britannica Films
425 N. Michigan Ave.
Chicago IL 60611
United States
800-554-9862; 312-347-7400

British Film Institute
21 Stephen St.
London W1P 1PL
Great Britain
1.255.1444

The Brooklyn Museum
200 Eastern Parkway
Brooklyn NY 11238
United States
718-638-5000
718-638-3731 (fax)

Brooks Institute of Photography
1321 Alameda Padre Serra
Santa Barbara CA 93108
United States
805-966-3888

Buffalo State College Film Library
1300 Elmwood Ave.
Communication Center 102
Buffalo NY 14222-1095
United States
716-878-6682

C.N.D.P.
Division des ventes
29 rue d'Ulm
Paris 75006
France
1.43.29.21.64

Madeleine Caillard
48 rue d'Hauteville
Paris 75010
France
1.47.70.03.62

Camera Three Productions
165 West Putnam Ave.
Greenwich CT 06830
United States
203-661-7500
203-661-1020 (fax)

Canadian Broadcasting Corp.
P.O. Box 500, Terminal A
Toronto Ont. M5W 1E6
Canada
416-533-3600; 416-925-3311; 416-975-6883
416-975-6887 (fax)

Canadian Filmmakers Distribution Centre
67A Portland St.
Toronto Ont. M5V 2M9
Canada
416-593-1808

Arthur Cantor, Inc.
2112 Broadway, Suite 400
New York NY 10023
United States
212-496-5710; 212-664-1298

Canyon Cinema Cooperative
2325 3rd St., Suite 338
San Francisco CA 94107
United States
415-626-2255

Carey Productions Ltd.
632 Ellengale Rd.
Burlington Ont. L7T 3N8
Canada
416-637-3025

Source Index

Carousel Films
260 Fifth Ave. Rm.705
New York NY 10001
United States
212-683-1660

Amon Carter Museum
P.O.Box 2365 - 3501 Camp Bowie Blvd.
Fort Worth TX 76107
United States
817-738-1933

CBC Enterprises
P.O. Box 500, Terminal A
Toronto Ontario M5W 1E6
Canada
416-533-3600; 416-975-3500; 416-925-3311; 416-975-7913

CEGEP de Trois-Rivières
3500 rue de Courval
Trois-Rivières Qué. G9A 5E6
Canada
819-376-1721

Center for Creative Photography
University of Arizona
Tucson AZ 85719
United States
602-626-4636

Centre Culturel Communal de Brétigny
rue Henri-Douard
Brétigny-sur-Orge 91220
France
1.60.84.38.68

Centre Culturel de Suède
11 rue Payenne
Paris 75003
France
1.42.71.82.20

Centre du Film sur l'Art
18 rue Joseph II
Brussels 1040
Belgium
2.217.28.92; 2.374.33.14

Centre Georges Pompidou
Musée national d'art moderne
Paris Cedex 04 75191
France
1.42.77.12.33

Centre Nat. de la recherche scientifique
CNRS Audiovisuel Serddav
1 place Aristide Briand
Menton Principal 92195
France
1.45.34.75.50

Centre National des Arts Plastiques
12 rue St. Florentin
Paris 75008
France
1.42.96.87.78; 1.42.96.68.56

Leonard Cheshire Foundation
Paddock Wood
East Knoyle
Salisbury Wiltshire
Great Britain
074.783.226

Christian Science Center
Film and Broadcasting Dept.
One Norway
Boston MA 02115
United States
617-262-2300

Churchill Films
12210 Nebraska Ave.
Los Angeles CA 90025
United States
800-334-7830; 213-207-6600
213-207-1330 (fax)

Ciné-Tamaris
86 rue Daguerre
Paris 75014
France
1.43.22.66.00; 1.47.34.57.17

Cinéculture
Ministére de la Culture et Comm.
3 rue Boissière
Paris 75016
France
1.47.27.30.60

Cinema Guild
1697 Broadway, #802
New York NY 10019
United States
212-246-5522
212-246-5525 (fax)

Cinéma Libre Inc.
3575 blvd. St. Laurent, bureau 704
Montreal Quebec H2X 2T7
Canada
514-849-7888

Cinémage
3 rue Saint-Augustin
Asnières 92600
France
1.47.93.11.11

Cinémathèque de l'Enseignement
Ministère de l'Education Nationale
B.P. 359
Montrouge Cedex 92541
France

Cinématographe Expansion
8 rue du 11 novembre
Sannois 95110
France
1.39.81.40.81

Cinq Continents
165 avenue de Clichy
Paris 75017
France
1.46.27.90.00

CNI Pictures, Inc.
53 Jean Dr.
Englewood Cliffs NJ 07632
United States
201-891-8240; 212-645-7522

Coast Telecourses
14460 Warner Ave.
Fountain Valley CA 92708-9527
United States
714-962-8861; 714-241-6109

Coll. Albert Kahn, Musée Départmental des Hauts-de-Seine
10, quai du Quatre-Septémbre
Boulogne 92100
France
1.46.04.52.80

Colstar Communications & Entertainment
1 Wardour Mews, D'Arblay St.
London W1V 3FF
Great Britain
1.437.5725

Concord Films Council Ltd.
201 Felixstone Road
Ipswich Suffolk IP3 9BJ
Great Britain

Consulate-General of the Netherlands
Press & Cult. Section
1 Rockefeller Plaza
New York NY 10020
United States
212-246-1429

Coronet/MTI Film & Video
108 Wilmot Rd.
Deerfield IL 60015
United States
800-621-2131; 312-940-1260

Pio Corradi
Zürcherstrasse 139a
Basel CH-8500
Switzerland

CRM Films
2233 Faraday, Suite F
Carlsbad, CA 92008
United States
800-421-0833

Czechoslovak Filmexport
Vaclacské nám 28
Praha 1 111 45
Czechoslovakia

Daedalus Productions
759 S. Sixth St.
Philadelphia PA 19147
United States

Fergus Davidson Associates
1 Bensham Lane
Croydon Surrey CR0 2RU
Great Britain
1.689.6824

DeCordova Museum and Sculpture Park
Sandy Pond Road
Lincoln MA 01773
United States
617-259-8355

Dee and Co. Ltd.
46 Potters Land
Barnet
London EN5 5BE
Great Britain
1.449.7533

Delta Image
225 boulevard MacDonald
Paris 75019
France
1.40.34.82.53

Direct Cinema Ltd.
P.O. Box 69799
Los Angeles CA 90069
United States
213-652-8000

Educational Foundation for Visual Aids
Nat. Audio-Visual Aids Library
The George Bldg., Normal College
Bangor Gwynedd L57 2PZ
Wales
248.355.155

Electronic Arts Intermix
536 Broadway, 9th Fl.
New York NY 10012
United States
212-966-4605

Embassy of France
4101 Reservoir Rd. NW
Washington DC
United States
202-944-6000

Encyclopaedia Britannica Educational Corp.
425 N. Michigan Ave.
Chicago IL 60611
United States
800-554-9862; 312-347-7400

Facets Multimedia
1517 W. Fullerton Ave.
Chicago IL 60614
United States
800-331-6197; 312-281-9075

FACSEA
972 Fifth Ave.
New York NY 10021
United States
212-439-1400

Fédération Européene des industries techniques du cinéma
50 avenue Marceau
Paris 75008
France
1.47.23.75.76

Film America, Inc.
1832 Biltmore St. NW
Washington DC 20009
United States
202-332-5817; 202-332-7032

Film Polski
Mazowieoka 6/8
Warsaw 00-048
Poland
48.26.34.17; 48.27.57.85

Film-Makers' Cooperative
175 Lexington Ave.
New York NY 10016
United States
212-889-3820

Filmakers Library
124 E. 40th St.
New York NY 10016
United States
212-808-4980
212-808-4983 (fax)

Les Films Armorial
233 bd Péreire
Paris 75017
France

Les Films Constance
20 passage de la bonne graine
Paris 75011
France

Les Films du Lagon Bleu
11 rue des Ursulines
Paris 75005
France
1.43.25.02.16

Les Films du Prisme
27 rue Louis-David
Paris 75016
France
1.45.03.42.82

Films for the Humanities
P.O. Box 2053
Princeton NJ 08540
United States
800-257-5126; 609-452-1128

Films Grain de Sable
206 rue de Charenton
Paris 75012
France
1.43.44.16.72

Films Inc.
5547 No. Ravenswood Ave.
Chicago IL 60640
United States
800-323-4222 x43; 312-878-2600

Les Films Jean-Noël Delamarre
c/o CAD Productions
36 rue Mathurin-Régnier
Paris 75015
France
1.43.06.58.95

First Run/Icarus Films
153 Waverly Place 6th floor
New York NY 10014
United States
800-876-1710; 212-727-1711

Glen Fleck Inc.
1526 Cloverfield Blvd.
Santa Monica CA 90404
United States
213-829-3567

Forma Art Associates Ltd.
125 E. 63rd St.
New York NY 10021
United States
212-371-9794

FR3
116 ave. du Président Kennedy
Paris 75786
France
1.42.30.22.12; 1.42.30.28.56

France Media International
78 quai Raymond Poincaré
Paris 75016
France
1.45.01.55.90

Hermine Freed
60 Gramercy Park N.
New York NY 10010
United States
212-475-5256; 212-924-2393

French Ministry of Foreign Affairs
Office of Audio-Visual Resources
21, bis Rue La Pérouse
Paris 75116
France
1.45.02.14.23

Friendship II Productions
11 51st St. C1
Weehawken NJ 07087
United States
201-866-9136

John Gati Film Effects, Inc.
881 7th Ave. Suite 832
New York NY 10019
United States
212-582-9060

Goethe Institut, Lille
Centre Culturel Allemand
98 rue des Stations
Lille 59800
France
20.57.02.44

Granada Television International Ltd.
Manchester 3 and Golden Square
London W1
Great Britain

Great Plains National Instructional Television Library
University of Nebraska
Box 80669
Lincoln NE 68501
United States
800-228-4630; 402-472-2007; 402-467-2502

Greenleaf Video Inc.
3230 Nebraska Ave.
Santa Monica CA 90404
United States
800-255-4687; 213-829-7675

Grove Press Films
920 Broadway
New York NY 10010
United States
212-529-3600

Guidance Associates
Communiations Park, Box 3000
Mt. Kisco NY 10549
United States
800-431-1242

Handel Film Corp.
8730 Sunset Blvd.
West Hollywood CA 90069
United States
213-657-8990

Heure Exquise Distribution
B.P. 113
Mons en Baroeul 59370
France
20.04.95.74

Home Vision
5547 No. Ravenswood
Chicago IL 60640
United States
800-826-3456 x325; 800-323-4222; 312-878-2600
312-878-8648 (fax)

HTV Limited
The Television Centre
Bath Road
Bristol BS4 3HG
Great Britain
222.59.0810; 272.77.8366

Gordon Hyatt Inc.
7 W. 81st St.
New York NY 10024
United States
212-874-0469
212-874-0509 (fax)

ICS
38 Avenue des Ternes
Paris 75017
France

ILEA Learning Resources Branch
275 Kennington Lane
London SE11 5QZ
Great Britain
1.633.5971

Images Productions
P.O. Box 46691
Cincinnati OH 45246
United States

INA
Relations extérieures
4 avenue de l'Europe
Bry sur Raine 75782
France
1.49.83.20.00; 1.43.47.64.00

Indiana University Audio Visual Center
Bloomington IN 47405-5901
United States
812-885-8087 (Sales); 812-885-2921 (Rental)

INF TV
95 boulevard St. Michel
Paris 75005
France
1.43.25.61.80

Inner-Tube Video
138 Prince St.
New York NY 10012
United States
212-966-7446

INRP
29 rue d'Ulm
Paris 75230
France
43.29.21.64

Institut d'Arts Visuels d'Orléans
2 rue de la Bibliothéque
Orléans 45000
France

Inter Ciné TV
9 rue Jean-Mermoz
Paris 75008
France
1.43.59.26.79; 1.40.74.00.66

Interama
301 W. 53rd St., Suite 19E
New York NY 10019
United States
212-977-4830

Intermédia
19 rue de Passy, 4e étage
Paris 75016
France
1.42.24.68.23

International Center of Photography
1130 Fifth Ave.
New York NY 10128
United States
212-860-1777

International Film Bureau
332 S. Michigan Ave.
Chicago IL 60604
United States
312-427-4545

Iskra
74 rue Albert
Paris 75013
France
1.45.83.11.18; 1.45.83.22.65

JRT/Yugoslavia Radio-TV
Generala Zdanova 28
Belgrade 11000
Yugoslavia
11.433.693

Friedrich Kappeler
Zürcherstrasse 139a
Basel CH-8500
Switzerland

Karol Media
22 Riverview Drive
Wayne NJ 07470
United States
201-628-9111

Kodak Video Programs
343 State St.
Rochester NY 14650
United States
716-724-1133

KS Visions
c/o Totem Production
27 rue de la Rochefoucault
Paris 75009
France
1.45.26.13.13

KUHT-TV
KUHT Film Productions
4513 Cullen Blvd.
Houston TX 77004
United States
713-748-6814

Gene Lasko
12 E. 86th St.
New York NY 10028
United States

David Latchman
147 Old Forest Hill Rd.
Toronto Ont. M5N 2N7
Canada
416-921-9527

Erwin Leiser Filmproduktion
P.O. Box 2112
Zurich CH 8028
Switzerland
1.251.52.89

Library of Congress
Motion Picture, Broad. and Rec. Sound Div.
Madison Bldg., Rm. 336
Washington DC 20540
United States
202-707-1000

Peter Lipskis Films
856 18th Ave. West
Vancouver BC V5Z 1W3
Canada
604-874-1486

London Film-Makers Co-operative
42 Gloucester Place
London NW1
Great Britain
1.586.4806

London International Film School
Dept. S27
25 Shelton St.
London WC2H 9HP
Great Britain
1.836.0826

Long Beach Museum of Art
2300 East Ocean Boulevard
Long Beach CA 90803
United States
213-439-2119; 213-439-0751

Lucerne Media
37 Ground Pine Rd.
Morris Plains NJ 07950
United States
800-341-2293; 201-538-1401

Magic Lantern Films
136 Cross Ave.
Oakville Ontario L6J 2W6
Canada
416-844-7216

Thierry Marzloff
64 rue de Silly
Boulogne Billancourt 92100
France
1.46.05.75.79

Mass Productions
233 E. 3rd St. #4C
New York NY 10009
United States
212-477-4309

Media West Home Video
10255 Southwest Artic Dr.
Beaverton OR 97005
United States
800-888-8273; 503-626-7002; 503-626-7004

Médiane
4 avenue Villars
Paris 75007
France
1.47.05.01.41

Médiathèque Centrale
Ministère des Relations Extérieures
38 avenue des Ternes
Paris 75017
France
1.45.72.21.10

MICA-TV
P.O.Box 217, Prince Street Station
New York NY 10012-004
United States
718-383-3287; 212-410-0882

Minneapolis Institute of Arts
2400 Third Ave S.
Minneapolis MN 55404
United States
612-870-3046

Ann Mundy and Associates
6726 Firelight
Dallas TX 75248
United States
214-392-7790

Musée d'Orsay
62 rue de Lille
Paris 75007
France
1.40.49.48.68; 1.46.33.78.14; 1.45.49.48.14
1.40.49.49.79 (fax)

Museum at Large, Ltd.
20 W. 22nd. St. #1402
New York NY 10010
United States
212-691-2977; 212-691-3220

Museum of Contemporary Art, Chicago
237 East Ontario St.
Chicago IL 60611
United States
312-280-2660

Museum of Fine Arts, Houston
P.O. Box 6826
1001 Bissonnet
Houston TX 77265
United States
713-526-1361; 713-639-7520; 713-639-7391

Museum of Modern Art Circ. Film Library
11 W. 53rd St.
New York NY 10019
United States
212-708-9530
212-708-9889 (fax)

Museum of Modern Art Dept. of Film
11 W. 53rd St.
New York NY 10019
United States
212-708-9400; 212-708-9614; 212-708-9530

Nat. Mus. of African Art/ Smithsonian Inst.
950 Independence Ave. S.W.
Washington DC 20560
United States
202-357-4600; 202-287-3490

National Audiovisual Center
8700 Edgeworth Dr.
Capitol Heights MD 20743-3701
United States
800-638-1300; 301-763-1896; 301-763-1860; 301-763-1850

National Film Board of Canada
P.O.Box 6100
Montreal Qué. H3C 3H5
Canada
514-283-9036; 514-283-9428; 514-283-9447; 514-283-9427; 514-333-3425

The National Gallery of Art
Fourth St. and Constitution Ave, NW
Washington DC 20565
United States
202-737-4215; 202-842-6249;
202-842-6353; 202-842-6273
202-842-2356; 202-289-5446 (fax)

National Trust Film Library
Unit 2, Cornwall Works
Cornwall Avenue, Finchley
London N31 LD
Great Britain
1.349.9155

New Day Films
853 Broadway, Suite 1210
New York NY 10003
United States
201-633-0212; 201-628-9111; 212-477-4604

New Yorker Films
16 W. 61st St.
New York NY 10023
United States
212-247-6110

Omnifilm Productions Ltd.
207 West Hastings St. Suite 1101
Vancouver B.C. V6B 1H7
Canada
604-681-6543

ORF
Programmadministration 2
Würzburggasse 30
Vienna A-1136
Austria
82.91.45.17

Pacific Arts Video
50 N. La Cienega Blvd. Suite 210
Beverly Hills CA 90211
United States
800-538-5856; 213-657-2233
213-657-0395 (fax)

Pacific Pictures
12 Hillcrest Ct.
Berkeley CA 94705
United States
415-547-3152

Paper Tiger Television
339 Lafayette St.
New York NY 10012
United States
212-420-9045

Pari Films
18 rue Vignon
Paris 75009
France
1.47.42.06.27

Paris Audiovisuel
14 rue du Colisée
Paris 75008
France
1.43.59.00.80

Past America Inc.
12100 N.E. 16th Ave.
Miami FL 33161
United States
305-893-1202

Edward Patterson Associates Ltd.
Treetops, Cannongate Rd.
Kent CT21 5PT
Great Britain
3.036.4195

PBS Video
1320 Braddock Pl.
Alexandria VA 22314
United States
800-424-7963; 703-739-5000; 212-753-7373

Pennsylvania State University Audio Visual Services
Special Services Bldg.
University Park PA 16802
United States
814-863-3106; 814-865-6314; 814-863-3102

Phoenix Films & Video
468 Park Ave. So.
New York NY 10016
United States
212-684-5910

Phoenix/BFA Films & Video
468 Park Ave. So.
New York NY 10016
United States
212-684-5910

Picture Start
221 E. Cullerton, 6th floor
Chicago IL 60616
United States
312-326-6233

Pinhole Productions
308 East 94 St. #2E
New York NY 10028
United States
212-831-8431

Polivideo S.A.
1 Via Stradonino
Riazzino 6595
Switzerland
41.92.640.111

Richard Price Television Associates
Seymore Mews House, Wigmore St.
London W1H 9PE
Great Britain

Pro Helvetia
Hirschengraben 22
Zurich CH-8024
Switzerland

Producer Services Group
7461 Beverly Blvd., PH
Los Angeles CA 90036
United States
213-937-5020

Productions de la Chartreuse
La Chartreuse
Villeneuve des Avignon 30400
France

Productions Lazennec
5 rue des Guillemites
Paris 75004
France
1.42.05.39.33; 1.42.77.61.86

Public Media, Inc.
5547 No. Ravenswood Ave.
Chicago IL 60640
United States

Pyramid Film & Video
Box 1048
Santa Monica CA 90406
United States
800-421-2304; 213-828-7577

Radiotelevisione della Svizzera Italiana
Casella Postale
Lugano CH-6903
Switzerland
91.58.51.11

Reed Vocational Resource Library
Reed College
Portland OR 97202
United States

Rental-see *Educational Film & Video Locator*, published by:
R.R. Bowker Co.
245 W. 17th St.
New York NY 10011
3rd Edition, 1986

Riff International Production
39 rue Censier
Paris 75005
France
1.47.07.58.22

Luciano Rigolini
Villa Gamine
Brusino Arsizio CH-6911
Switzerland
91.69.13.56; 91.71.53.67

RM Associates/NY
250 W. 57th St., Suite 1005/6
New York NY 10019
United States
212-262-3230
212-262-5146 (fax)

Rodman-Downs, Ltd., Inc.
75 Rockefeller Plaza
New York NY 10019
United States
212-246-8200

The Roland Collection
3120 Pawtucket Rd.
Northbrook IL 60062
United States
312-291-2230

San Francisco Center for Visual Studies
49 Rivoli St.
San Francisco CA 94117
United States
415-664-4699

San Francisco Production Group
550 Bryant St.
San Francisco CA 94107
United States
415-495-5595

SBS Television Marketing
P.O. Box 028
4 Cliff St.
Milson's Point NSW 2061
Australia
2.964.2828
2.957.3571 (fax)

La Sept
35 quai André Citroën
Paris 75015
France
1.40.59.39.77
1.45.78.09.27 (fax)

Show Corp.
79 Victory St.
Stamford CT 06907
United States

Silver Productions
260 Beacon St.
Boston MA 02116
United States
617-266-6482

Smithsonian Institution
National History Bldg., Rm. C222B
10th & Constitution Ave. NW
Washington DC 20560
United States
202-357-2984

Smithsonian Institution Office of Museum Programs
2235 A & I Bldg.
Washington DC 20560
United States
202-357-3101

Société Française de Production
36 rue des Alouettes
Paris 75019
France
1.40.03.50.00

Société Nouvelle Pathé Cinéma
6 rue Francoeur
Paris 75018
France
1.42.57.12.10

SODEP
19/21, rue Rasson
Brussels 1040
Belgium
2.736.5437

Cecile Starr
50 W. 96th St.
New York NY 10025
United States
212-749-1250

John A. Stewart Productions
No address known
United States

Syracuse University Film Rental Center
1455 E. Colvin
Syracuse NY 13210
United States
315-443-2452

Tom Taplin
715 Ashland Ave.
Santa Monica CA 90405
United States
213-396-1178

Katherine Deutch Tatlock
157 Naples Rd.
Brookline MA 02146
United States
617-277-5040

Team Video Productions
Canalot, 222 Kensal Rd.
London W10 5BN
Great Britain
1.960.5536

Television of the German Democratic Republic
East Berlin 1199
German Democratic Republic
2.631.5880

TF 1
Service Commercial
17 rue de l'Arrivée
Paris 75015
France
1.42.75.12.34

Turner Broadcasting Services/WTBS
1050 Techwood Dr. NW
Atlanta GA 30348
United States
404-827-2169

Turner Home Entertainment
5 Penn Plaza, 19th Fl.
New York NY 10001
United States
212-714-7976; 212-714-7975
212-714-7801 (fax)

Katherine Tweedie
3911 rue de Bullion
Montréal Québec H2W 2E2
Canada
514-849-1043; 514-848-4605
514-848-8627 (fax)

Tybanis Film
223 bd. Pereire
Paris 75017
France
45.74.72.72

UGC Droits Audiovisuels
2405 avenue Charles de Gaulle
Neuilly-sur-Seine Cedex 92200
France
1.46.40.44.36
1.46.40.44.30 (fax)

United States Information Agency
400 Sixth St. SW Rm. 567
Washington DC 20547
United States
202-485-2783; 202-376-2116; 212-655-4000

Universidad Nacional Autonoma de Mexico
Coord. de Extension Universiteria
Mexico City
Mexico

University of Illinois Film Center
1325 S. Oak St.
Champaign IL 61820
United States
217-333-1360

University of Southern California Film & Video Distr. Ctr.
University Park
Los Angeles CA 90089-2211
United States
213-743-2235

University of Toronto/Media Centre
121 St. George St.
Toronto Ont. M5S 1A1
Canada
416-978-6049
416-978-7705 (fax)

Varied Directions
69 Elm St.
Camden ME 04843
United States
800-888-5236; 207-236-8506
207-236-4512 (fax)

VidAmerica Inc.
235 E. 55th St.
New York NY 10022
United States
212-355-1600

Video Data Bank
School of the Art Institute of Chicago
Columbus Drive at Jackson Blvd.
Chicago IL 60603
United States
312-443-3793

Video Gallery
No address known

Walker Art Center
Vineland Place
Minneapolis MN 55403
United States
612-375-7600; 612-375-7621
612-375-7618 (fax)

Wallonie Bruxelles Images
rue des Mimines 26
Brussels 1000
Belgium
2.518.1460

WDR International
Postfach 101950
Köln 1 D-5000
Federal Republic of Germany
221.220.2775

Westdeutscher Rundfunk
Appellhofplatz 1
Köln 1 5000
Federal Republic of Germany
221.220.2955; 221.221.2201
221.220.4800 (fax)

WETA-TV
Box 2626
Washington DC 20013
United States
703-998-2600; 703-998-2709; 703-998-2833; 703-998-2713

WNED-TV
184 Barton St.
Buffalo NY 14213
United States
716-881-5000

WNYC
1 Centre St. 26th Floor
New York NY 10007
United States
212-669-7700

Wombat Film & Video
250 W. 57th St., Suite 2421
New York NY 10019
United States
212-315-2502; 800-542-5554 (exc. NY);
212-582-2585; 212-582-0585 (fax)

WTVS
7441 Second Blvd.
Detroit MI 48202
United States
313-873-7200, ext. 277

Yellow Cat Productions
300 Eye St. NE, Suite 1
Washington DC 20002
United States
202-543-2221

ZDF
Sales Department
Postfach 4040
Mainz 1 D-6500
Federal Republic of Germany
6131.702.080

Zipporah Films, Inc.
One Richdale Ave., Unit 4
Cambridge MA 02140
United States
617-354-2277